T0274091

IN THE TREACLE MINE

THE LIFE OF A MARINE ENGINEER

J.W. RICHARDSON

Whittles Publishing

Published by
Whittles Publishing,
Dunbeath,
Caithness KW6 6EG,
Scotland, UK

www.whittlespublishing.com

ISBN 978-184995-488-4

Printed in the UK by Page Bros (Norwich) Ltd

By the same author
'Over the Alps' on the Watercress Line

CONTENTS

INtRODUCtION

This book is about the 29 years I spent in the Merchant Navy, starting in 1964 as an engineer cadet with Esso Tankers and retiring as chief engineer on the cross-Channel ferries in 1993. There are many other maritime biographies to be found, but nearly all of them have been written from the point of view of the captain or sailors, and it can easily be forgotten that on any ship without sails at least a third of the crew will be engineer officers and ratings (I am not counting cruise ships and the like, where the cabin and catering crew outnumber everyone else put together). This book, then, attempts to redress some of this inequality and to show that the members of the Black Gang also have a tale to tell.

Many books and films appear to be completely oblivious to the part played by the engineers on board ship, especially during wartime, when they were just as likely to die as anyone on the bridge. Although the captain was indeed the man who had to decide whether to turn and fight the German pocket battleship or to make smoke and run for it, he would not get very far without his engineers to provide the power (or indeed the smoke) to do it.

I would hasten to add at this point that I am not trying to stir up the old oil and water controversy, which dated back to the days when some ships still had sails as well as the new-fangled engines. Back then the engineers had to endure some resentment from certain captains and deck officers, who might have felt that the advance of technology had taken away some of their authority, and also perhaps because their pristine holystoned decks were now occasionally blackened by soot and coal dust. I can truly say that during my time at sea I never experienced any animosity on the part of the navigating officers toward the engineers simply because they were engineers, or vice versa.

This book is not attempting to be a treatise on marine engineering, either, although it would be pointless to write a book from the perspective of an engineer without some descriptions at least of the machinery and how it works, in the same way that a deck officer would want to talk about the intricacies of navigation, working cargo and ship handling. I have endeavoured to spread this engineering speak evenly around the book, so that you won't feel too overloaded by too many technicalities being thrown at you in one go, although the chapters dealing with my first voyages have a rather bigger dose, in order to provide enough background information to make the subsequent ones easier to follow. I have described some of the work done by the engineers in detail, so you will realise that there is rather more to the job than just pulling levers and pressing a few buttons.

The main purpose of the book, however, is simply to paint a picture of my life at sea, the places I saw and some of the characters who shared it with me, mostly for the benefit of my family but also for anyone else with maritime interests. It may also appeal to the growing band of steam preservation enthusiasts. Although most of my steamships were turbine-driven, one whole chapter is devoted to a ship with a steam reciprocating engine, and there are numerous other mentions of this type of machinery.

Regrettably, as this story starts some 50 years ago and I did not keep a diary, I have forgotten the names of most of my shipmates, especially from the early years. In certain cases, however, where I can remember them, I have just used their initials or their rank, to help maintain their privacy and that of their relatives.

The other consequence of my failing memory is that although the events I have described are all true and as accurate as I can make them, they may not be in strict chronological order. Apart from the years 1974 to 1978, when I have some of my wife Astrid's letters to her parents to provide details of where we went, most of the other dates and ship details have come from my discharge book and my own memory.

When I started work on the ferries not even the discharge book was a completely reliable guide, as it was common practice for crew members to swap about between company ships to cover leave or sickness when required. During my time with Sealink/Stena Ferries it became even more difficult, as the articles were signed in the company office in Dover instead of on board ship, and were worded to cover service on any company ship rather than a specific one.

I believe, however, that the story itself is the important thing, rather than knowing the exact time and place, especially when nearly all the ships concerned have already made their final journey to the scrapyard. The book, then, is simply a collection of purely personal reminiscences, the main thrust of which is to show what it was like to be a marine engineer starting his career in the 1960s when the steam age was coming to an end and the all-conquering diesel engine was taking over.

Throughout the book I have used imperial units for the most part because that is what I was brought up with – besides which, an engine room temperature of 100°F sounds a lot more impressive than 38°C! In case anyone still remains confused, I have included a glossary with a few conversions in it, although most things are explained in the text as I come to them, which should save you from having to refer to it too often.

John Richardson, August 2021

ACKNOWLEDGEMENTS

I would like to thank the following people who have helped in the production of this book. Firstly my wife Astrid, who apart from anything else features quite a lot in these pages and who has always been a wonderful companion and shipmate. She was also able to provide details of some of the places we visited, as some of the letters she wrote home to her parents during her time at sea with me have survived.

I would also like to mention some of my former shipmates who have shared the story with me: Rob Caldwell, Mike Forwood, Tony Warne and Norman Axford, in particular, who I am still in contact with and who have helped fill in a few gaps when my memory failed me.

For the rest, I am afraid I have forgotten nearly all the names, although the faces may remain clear, so should any of them read this and fail to find their name listed, please accept my apologies. Nevertheless, here are a few: John Archdeacon, John Armstrong, Peter Brinkley, Peter Button, Dave Cartmer, Michael Champneys, the 'real' Dave Clark, Nigel Cureton, Chris Davison, Jim Duncan, Ron Ellison, Neil Farquhar, Dick Gammie, Dave Giddy, Clive Godderidge, Graham Hale, John Hayes, Steve Humphries, Pat Lockhart, John McNeil, John Newall, Norman Oakley, John Pett, Frank Quick, Martin Roberts, Steve Rooke, Brian Smith, Geoff Waddington and 'Uncle' Ernie Ward.

The Institute of Marine Engineers (now ImarEST) also deserves a mention, firstly for giving me a grant to study at university and also for its excellent monthly publication, *The Marine Engineers Review*, which attempted (occasionally successfully) to keep me up to speed with the advance of modern maritime developments.

Thanks in particular to my proofreaders, Mark Rudall, Mike Smith and Mike Bell.

Thanks too to Google and Wikipedia, which between them provided a lot of the facts and figures quoted in this book.

Finally, especial thanks to Nigel Thornton of the Dover Ferry Photos website, who obtained the required permission to publish the photos of MV *Tiger*, MV *Hengist* and *Hoverspeed Great Britain*, and to the website itself, which provided details of all the ferries I sailed on.

GLOSSARY AND ABBREVIATIONS

AC	Alternating current, also air conditioning.
Aft	Toward the stern (back end) of a ship.
BA	Breathing apparatus.
B&W	Burmeister & Wain – a make of marine diesel engine.
Bandhari	An Indian cook.
Board of Trade sports	Lifeboat drill.
Bond	The bonded store, where all the booze and cigarettes are kept locked up when in port.
bhp	Brake horse power
Blower	On motor ships a turbocharger; also an engine room ventilator outlet.
BSP	British Standard Pipe thread.
Bottom end	The bearing at the foot of the connecting rod, where it joins the crankshaft – like a big end in a car.
Box up	Engineers' expression meaning to reassemble.
Bows	The front of the ship.
Bulkhead	What you would call a wall in a building – on a ship it could range from a simple partition between cabins to an extensive heavy steel one dividing the ship into watertight compartments.
Bumboat	Like a travelling market stall, selling goods to ships' crews.
Butterworth	Trade name of the tank cleaning machine, also applied to the pump and heater (when used) that supplies it.
CPP	Controllable-pitch propeller.
Capstan	A winch used for pulling in the mooring ropes – usually with drum mounted vertically.
Chronometer	The very accurate ship's clock, kept in the chart room, that was required for navigational purposes.
Condenser	A heat exchanger that turns steam back into water – the main one on a steamship can be very large and uses sea water for cooling.
Crosshead	A bearing between the end of the piston rod and the top of the connecting rod on large two-stroke diesel engines and steam reciprocating engines.

Davits	A kind of crane: consisting of two jib arms and a winch, by means of which the lifeboats can be swung out and lowered.
De-aerator	Part of the boiler feed water system – apart from getting rid of any air, it also acts as a heater and as a reservoir of feed water and is usually placed high up in the engine room.
Deckhead	What you would call the ceiling in a building.
Drop the pick	Lower the anchor.
dwt	Deadweight tons (a measure of a tanker's carrying capacity).
Economiser	A boiler feed water heater that uses the hot exhaust gas from the boilers as the heating medium.
Evaporator	Device on board ship for making distilled water from sea water, for drinking and boiler use.
Fahrenheit	Imperial unit of temperature – to convert to Celsius, subtract 32 and multiply by 5/9.
Feed pump	The pump that supplies water to the boilers.
Feed water	The water used to supply the boilers.
Forecastle	The interior spaces at the bows of the ship – on old sailing ships it was where the crew's living quarters were situated.
Forecastle head	The open deck at the bows where the windlasses are to be found.
Heel	When a ship leans over to one side – usually just a temporary state caused by wind, wave action or when making a sharp turn.
HP	High pressure.
Handomatic	Engineers' term meaning manually operated.
Have a blow	Take a break
Galley	The ship's kitchen.
Genny	The usual term for any machine on board ship that generates electricity.
Get paid off	Sign off articles and leave the ship, when the departing crew would get their wage packets. The term is still used today even though salaries are now usually paid directly into the bank.
Gland	That part of a pump or valve where the shaft or spindle emerges – it will be sealed either by a stuffing box (see below) or a mechanical seal.
GPS	Global positioning by satellite (satnav).
grt	Gross registered tons (the carrying capacity of a cargo ship).
Gripes	A pair of wire hawsers locked in place with quick-release hooks, which secure the lifeboat in the davits.
LP	Low pressure.
Lecky	The ship's electrical officer.
List	When a ship has a permanent heel to one side.

MWM	A make of marine diesel engine.
nm	Nautical mile – 6,080 feet; or one minute of arc at the equator.
Plates (the)	The engine control platform.
ppm	Parts per million.
psi	Pounds per square inch – the imperial unit of pressure; 14.5 psi is approximately 1 bar.
Red Duster	The British Merchant Navy flag – more correctly the red ensign.
RIB	Rigid inflatable boat.
rpm	Revolutions per minute.
Sextant	A device used in navigation for measuring the angle between a celestial body and the horizon.
Scavenge port	The air inlet port on a marine diesel.
Scavenge trunking	The air inlet manifold on a marine diesel.
shp	Shaft horse power.
Sparks	The ship's radio officer.
Stuffing box	A recess around a shaft or spindle, into which packing can be inserted to form a seal, the packing being held in tight by a gland piece, plus studs and nuts.
Superheat	A condition of steam when its temperature is raised above that of the boiling water it came from. At normal atmospheric pressure water boils at 100°C, and if the steam is heated above this, then it is said to be superheated. In Esso, where the boilers on some ships worked at 860 psi (60 bar), water boils at 275°C and was superheated to 500°C.
Superheater	The device that does the superheating – basically by passing steam from the boiler drum back through another set of tubes within the furnace spaces.
Tail shaft	The final section of the propeller shaft, with the propeller fixed to it. It extends from the propeller into the shaft tunnel compartment, where it will be bolted onto the next section.
Tank top	The inner bottom of a double-bottomed compartment.
Telemotor steering	A system whereby turning the ship's wheel sends hydraulic oil via small-bore pipes to and from the steering flat to operate a hydraulic servo. This in turn signals the main steering pumps to move the rudder.
Turbo	On a steamship it means driven by a steam turbine – could be a turbo alternator for producing electricity or a turbo pump, cargo pumps and feed pumps being the most common ones. On motor ships it will mean the turbocharger: this consists of a turbine driven by the exhaust gas connected to a compressor that provides combustion air for the engine.

Ullage cap	A small opening in a tank lid through which a sounding tape can be inserted to measure the contents – it is usual to take an 'ullage' in a fuel or cargo tank, which is the depth to the top of the liquid, as opposed to a 'dip' which is the measurement to the bottom of the tank.
UMS	Unmanned machinery spaces.
Up and downer	A steam reciprocating engine.
Uptakes	The exhaust pipes (for both steam and diesel ships) that lead up into the funnel.
VLCC	Very large crude carrier – 100,000 tons and over.
Watch	A four-hourly period of duty.
Windlass	A winch on the forecastle head with a horizontal shaft, used for heaving up the anchors.

FEELING THE HEAT

I have read that for many years El Azizia in Libya held the record for being the hottest place on earth, when the temperature there once reached 136°F (58°C) in the shade. In 1969 I was an engineer officer on an oil tanker called the SS Esso *Durham* out in Bonny, Nigeria, where our average engine room temperature was always the wrong side of 120°F at the control platform, and in the boiler room it was much higher. At the bottom by the furnace fronts, where there was plenty of ventilation, it was often over 130°F (and this, strangely enough, often felt better to work in than the engine room). Right at the top, by the uptakes, it was usually around 150°F, and on one occasion it reached 156°F (69°C). I know these figures are true because every watch I had to climb up two flights of steps to the thermometer to record them for the engine room log, and they were so exceptionally high that I have never forgotten them.

Even that was not the hottest place where you could go in the ship: at this top level there was a walkway that actually passed between the two boiler uptakes, where of course there was a lot of radiated heat as well. The only reason to be there normally was to take flue gas samples for analysis – fortunately this didn't happen very often on *Durham*. Just for a change, I once hung the thermometer on the handrail there instead to see what it would get up to – it reached an astounding 168°F (76°C). That remains the all-time hottest place I experienced during my career at sea, and one that could really only be endured for the time it took to go in and read the thermometer.

Metal at these temperatures will burn the skin, and watches, bracelets, St Christopher chains etc cannot be worn. Here is an example that I think helps prove the point. The engineers' alarm sounded off one day when I was in my bunk, and in my hurry to get down below I didn't put on any socks. I soon noticed that the tops of my feet were starting to get sore and when the panic was all over and I got back to my cabin, I found that each of my feet had two rows of red marks on top where they had been burnt by the brass lace eyelets of my boots.

How we stood the gruelling long hours of hard physical work and rapid decision making at these temperatures, was basically by drinking fluids by the bucketful and taking so much salt that any nutritionist sitting in his nice cool office would be horrified. *Why* we stood it, when any self-respecting shop steward would be leading his troops out of the factory if the temperature got much above 85°F (29°C), may need some explanation. In the first place we didn't have any shop stewards. Having signed articles, we were under the captain's orders, so any refusal to work would be mutiny. Naturally, we could no longer be flogged, but going on strike would still be a serious offence and would no doubt mean we would never get another job at sea. The main reason, however, is nothing to do with rules and regulations but simply that we were all (literally) in the same boat – if we lost our engine and the ship was wrecked, then we would all be in equal peril. There was also a certain *esprit de corps* among the engineers in particular, which meant that we stuck together and tried to make the best of it. Finally, this was a very well-paid job, which we were unlikely to be able to match ashore.

How, you might wonder, did I end up in a situation like this? Read on …

1 LEAVING SCHOOL

When I was a little boy, I wanted to be an engine driver. This was, of course, back in the days when trains were still mostly hauled around by steam locomotives. My youthful ambition lasted right through into my teens, when it had become obvious even to me that by the time I left school there weren't going to be any steam engines left for me to drive – and additionally (thanks in large measure to the infamous Dr Beeching), there were going to be precious few railway lines remaining open anyway.

Gradually, as these unpleasant facts began to sink in, my thoughts were inevitably forced into considering an alternative career. I was good with my hands and very much enjoyed tinkering about in the garage at home, helping my dad make bits and pieces of darkroom equipment (he was a very enthusiastic amateur photographer), or my elder brother Peter, who usually had one or two motorbikes in various states of disassembly. This ability, and the fact that I detested the thought of working in an office all day, led me to start thinking about a career in mechanical engineering.

I was still reluctant to completely give up the idea of working on the railways and found that British Railways itself offered a number of opportunities for entry into the profession, either by way of a traditional five-year craft apprenticeship or else as graduate entrant, for which it paid something in the way of a bursary while you were at university. I even went as far as to get enrolled on a three-day introductory course based up in London, where the prospective applicants attended various presentations in the mornings and then in the afternoons went off by train to visit various railway workshops and engine sheds. The works visits, which included Swindon (home of the Great Western Railway), I found most interesting, of course, as there were still a few steam engines around to be seen – but the general impression I got was that the whole system was very tired and the workforce disillusioned (as well they might be), and that it was not where I wanted to be for the rest of my life.

At this time I was in what was to be my final year at school – I had been persuaded that if I wanted to do anything in engineering, then the only way to go was via a university degree. For entry onto an engineering degree course, I was supposed to need at least three GCE A levels at grade B or above in maths and physics, plus one other subject (in fact any three sciences would have done). My three best subjects were biology, chemistry and geography, but these apparently did not suit the school timetable and I was more or less bullied into doing pure and applied maths and physics instead. These turned out to be very poor choices and I struggled from day one – I can remember a test paper we had once where I scored a measly 4 per cent in pure maths!

If truth be known, I had by that time already decided to leave school at the end of term anyway, without carrying on for another year and actually sitting the A-level exam, which I am sure my school was more than happy about. Many people think that school league tables are a relatively modern idea, so it may come as a surprise to them that even as far back as the sixties schools were still very aware of how they compared with their neighbours. I am sure, therefore, that my school believed it would be better for their figures if my results weren't part of them. Even for O levels (I got seven of these), they cherry-picked the papers that each student was allowed to sit: in my case, even after studying history and French for five years, I was still not considered good enough to be allowed to sit those exams – a decision which rankles with me to this day.

Regarding careers advice at school, there was precious little on offer. I had one ten-minute interview with a school leaving officer, who seemed to think that I hadn't done my homework on what options might be available to me (he was quite right), and was therefore not inclined to waste any more of his precious time on me. At one stage he asked me quite gruffly if I even knew what a CAT was (College of Advanced Technology) – which had me stumped, and my attempt at a humorous reply (four-legged furry thing; purrs when stroked) left him singularly unmoved. Following this unhappy meeting I was left entirely to my own devices, with the suggestion that I had better go to the school library where there were a few books and leaflets that might give me some pointers.

I duly spent a couple of lunchtimes in the library where there was one quite thick tome listing just about every possible form of employment from coal mining to midwifery and how to get into it. I was idly leafing through this when a section on the Merchant Navy caught my eye. At a time when £1,000 a year was considered to be a good income, I saw that sea captains could earn three times that amount and that chief engineers got almost as much. I had never really thought about a career at sea, although my half-brother Peter, who had done his National Service in the Royal Navy, had obviously had a jolly good time. He had been the captain's secretary on HMS *Ceylon*, which was the Colony class cruiser that had accompanied the Queen and Prince Philip when they did a world tour on SS *Gothic* in 1952–3, and had seen quite a bit of the world (the bits that were coloured pink on the map, that is!). Furthermore, the qualifications required to embark upon a sea cadetship were simply an O level in maths and English if you wanted to go in on the deck side (leading eventually to captain), or four O levels, which had to include maths, English and one science subject, for the engineers.

My love of things mechanical and the knowledge that although steam was fast disappearing from the railways there were still plenty of steamships about made the decision whether or not to go on the deck or engine side an easy one to make, so I applied to a number of different shipping companies for details on their engineering cadetships. Many times over the years to come, when I was sweating buckets down in the bowels of some ship fixing the engines while I knew the deck officers would be simply lounging around on the bridge waiting, I was to wonder whether or not I had made the wrong decision.

After a few days, several glossy brochures arrived through the post, which, apart from giving details of the form the training would take, also included quite a few pictures of the various ships and their happy, smiling crews. The cadetship was split up into three separate sections: the first two years were spent entirely in college studying for an OND (Ordinary National Diploma) in mechanical and electrical engineering. Then there would be 18 months at sea on the company ships, and finally a year in an engineering workshop, gaining fitting and machining experience. Some companies did the workshop phase second and the seagoing experience last; this was called a reverse phase cadetship. I sent off applications to British India Steam Navigation Co. which at the time had well over 100 ships in service, and Esso Tankers (the British division) which had around 30 ocean-going ships and a fairly large coastal fleet as well. ('Esso', incidentally, is an acronym for Standard Oil (ess-oh) of New Jersey, the American parent company.)

Quite why one of my choices was a tanker company I now find rather difficult to explain, as it was well known that tanker men only spent a fraction of the time in port that might be spent by their contemporaries on a general cargo ship. Possibly it was because the tanker crews worked shorter periods of duty before being relieved – typically four-month trips instead of six to nine for general cargo (like British India), and as the only child of my parents' union I knew it would be a big wrench for them that I was going away at all. Additionally, the pay was higher on tankers and the promotion prospects much more rapid, so it could have been that I fancied getting rich quick rather than spending all my money on runs ashore. Anyway, the die was cast; I had been invited for interviews by both of my chosen companies, and unless both of them offered me a job my mind would be made up for me.

My first interview was with the British India Line at its traditional old-fashioned offices in Leadenhall Street in the East End of London. The main memory I have of the place was the magnificent models of the company ships that were spread around the building – each one a real masterpiece, with every derrick and pulley block faithfully reproduced; even the portholes had brass rims with glass inserts. As to the interview itself I can remember very little, except that part of it consisted of a kind of psychological profile test (quite advanced thinking for those days), in which I was presented with a piece of paper with about 40 words on it, each of which had a double meaning. I was supposed to go quickly through it, ticking the meanings I first thought of. One of the words was 'sink' and the choices were 'wash hand-basin' or 'opposite of float'. I chose the latter. When British India replied rejecting my services (and it was a beautifully written and polite letter, wishing me all the best in my future endeavours), I often wondered whether it was my pessimistic definition of the word 'sink' that had something to do with it.

Anyway, there was no time to brood on it as a few days later I was off again to London to see Esso. Its offices were in Victoria Street on the other side of town, and were much more modern – all concrete, glass and plush carpets. There was just one model ship on display, however, but it too was rather fine. I can't recall anything much about this interview either, although I vividly remember the medical that followed, during which I had the novel experience of having my balls gently cupped by the woman doctor whilst being asked to cough. All must have gone well, however, because within a few days I had the formal offer of an engineer cadetship with Esso Tankers, starting in September 1964, with a salary of £220 a year plus 7 shillings a day food and travel expenses (today worth about £4,600 + £7).

So I was going to be a marine engineer and not an engine driver after all. As it happened, some 30 years later I became involved with a railway preservation society in my spare time and was eventually passed out as a steam locomotive driver in 2004. If you want to know more about that, then you will have to read my book '*Over the Alps' on the Watercress Line*, published in 2008.

There was now just one hurdle to be surmounted: an interview by the principal at Poplar Technical College, where I would be spending the next two years while studying for my OND. The interview was really a bit of a formality, as the college existed on the fees paid to it by the shipping companies, and having agreed to provide a certain number of places it was highly unlikely to turn away anyone who had passed the company interview. The principal was one Professor G, who for some reason took an instant dislike to me. One of the things he asked me was whether I could explain what the four strokes in a four-stroke engine were, to which I replied 'Suck, squeeze, bang, blow,' meaning induction, compression, power and exhaust strokes. This, he said, was a frivolous reply and told me that I needed to develop a more serious attitude if I was going to get anywhere in the profession. As time went by, in common with nearly all my fellow cadets, I found him to be a completely humourless individual who would treat even the slightest misdemeanour as if it was a hanging offence; he remains almost the only person I ever met during my career that I really disliked.

The interviews had taken place sometime around June, so I still had a few weeks left at school to put up with. These passed dismally slowly, as I no longer cared about A levels and had stopped doing any real work – I spent most of my time looking out of the windows and daydreaming about a life on the ocean wave. On one occasion during a physics lesson I was rudely disturbed from one of these reveries by the teacher asking me why (with reference to Newton's third law of motion) it would be impossible for me to lift myself off the ground by pulling up on my bootlaces. The correct answer was of course, that for every action there is an equal and opposite reaction, which in this case would cancel each other out so I would remain firmly rooted to the floor. However, as I was no longer afraid of the consequences of a cheeky reply, I declared that I didn't suppose my bootlaces were strong enough, which caused considerable mirth amongst my classmates and even raised a (rather exasperated) smile from the teacher. Our unfortunate instructor then posed the question to one of the brighter sparks in the class, who could normally be relied upon to give the right answer. This chap must also have succumbed to a fit of devilment, because he replied, 'I couldn't do it either, sir, because I have elastic-sided boots with no laces' – this of course had us all in hysterics.

Strangely enough, there were no unpleasant repercussions from this little episode, despite it being a school where discipline was rigidly enforced (I was once caned for drawing a small cartoon figure on a document pinned up on the school noticeboard), and from then on it seemed to be accepted that as I was leaving anyway there was not much point in telling me off for not doing my homework, or indeed any other work for that matter.

Our school put sporting prowess very high up on the list of desirable qualities, and as I detested games I was never held in much regard by most of the teaching staff. The one exception was Mr Field, who took us for English and who always seemed to be interested in what I was doing; knowing of my fascination with steam trains, he nicknamed me 'engine driver'. He always wore his academic gown during lessons and sometimes even his mortarboard as well, and had the habit of walking up and down amongst the class and swatting pupils round the ear with one of the tails of this gown, into which he had tied a knot for the purpose, if he thought they were not paying attention. This was never done with any apparent malice even though it did in fact hurt quite a lot. Anyhow, I very much liked Mr Field. At the end of morning assembly he would occasionally play the school organ, which was quite a fine instrument with two manuals and dozens of stops, but unlike the music master himself, who would usually play some rather dreary work, Mr Field would belt out Widor's Toccata or some other rousing piece. I am afraid I never shone at his subject although I did get a C at O level, which was at least respectable, and I am sure that if it had been suggested to him that one day I would actually have a book published, he would have been amazed but hopefully a little proud as well.

At one time it was common for parents to say to their children that that they should make the most of being at school because 'schooldays are the best time of your life'. I never found this to be true and couldn't wait for the end of my final term. When it eventually arrived, in July 1964, I was only too happy to step outside the school gates for the last time.

2 POPLAR tECH

When Esso offered me the cadetship I was given the choice of three colleges for Phase 1 of the training. I could have gone to Southampton or South Shields or Poplar in the East End of London. I chose Poplar, as it was nearest to my home and meant that I could stay on with my parents. Opened in 1906, the college, which is quite an imposing building (now Grade 2 listed), stands on the south side of Poplar High Street, which runs parallel to the Commercial Road. Looking out from the windows to the rear of the building where most of the classrooms were situated, you could see some railway lines and beyond them what appeared to be a 30-foot brick wall extending hundreds of yards in each direction; this was in fact the backs of the warehouses which lined the quays of the West India Docks, and the masts and funnels of the ships therein could be seen poking out above it.

Poplar in general was still a very rundown area at this time, having been heavily bombed during World War II, but the college appeared to have come through unscathed, although it had been somewhat disfigured at a later date by scruffy additions to the sides and rear. In the entrance lobby stood a very fine model of a Doxford two-stroke opposed-piston marine engine, which could be made to run at the press of a button. This, then, would be where I would be spending the next two years, and I have to say that in common with the majority of my fellow cadets (the annual intake was about 50 split into two classes), I was not very impressed – especially after the introductory pep talk by Professor G, which consisted mostly of explaining all the college rules and the dire consequences of breaking any of them.

As the cadets were drawn from all over the country, most of them had to find lodgings in the surrounding area, and quite a few fetched up in the Queen Victoria's Seamen's Rest on the Commercial Road. A minority, however, including me, commuted in every day, and it turned out that this was a real bind. At the time I was living in Shirley, a suburb halfway between Bromley and Croydon on the south side of London, and I was faced with a 30-minute walk to West Wickham station, where I would catch a train to New Cross. From there it was a tube ride under the Thames (using Marc Isambard Brunel's original tunnel!) to Shadwell, then another short walk down onto the Commercial Road, and finally a bus up to Poplar. To get to college for 9 am, I had to leave home at 7 am and as classes didn't finish until 5 pm, it meant I was seldom home until gone 7 in the evening, after which there would always be a couple of hours of homework. I had never worked particularly hard at school, especially during my final year, and it came as quite a shock to the system to be faced with this unrelenting routine.

We were tested at regular intervals all through the course and if we were found wanting to any degree it would mean an unpleasant visit to the principal's office, where we would be read the riot act and threatened with various punishments, starting off with being reported to our company and finally with dismissal. The college prided itself on achieving at least a 95 per cent pass rate for the OND exams, and there didn't appear to be any fall-back plan in the event of a failure: retakes were not allowed except in extreme cases, such as serious illness or accident. Two or three cadets dropped out during the first year (accounting for the 5 per cent odd), after which the remainder were all expected to pass – and we all did.

Some of the lecturers were pure academics who had never worked at anything other than teaching, but quite a few were ex-marine engineers who had decided for some reason to 'swallow the anchor' and work ashore. In the former group was one Dr S, who took us for thermodynamics and heat engines – a deadly boring subject if ever there was one. He was always rather shabbily dressed and inevitably wore leather sandals which slapped a bit as he walked and gave us a little advance warning of his approach – useful if he had ever left us unattended in class and we were up to any sort of mischief, like simply talking amongst ourselves. He had a particularly vile temper and would become almost apoplectic with rage if we ever seemed to be giving one of his dull presentations anything less than 100 per cent attention. In my view this made him the worst type of teacher, because we were all too afraid of him to ask a question and so if we needed any help we had to rely upon each other after the class. It was harder still in my case, because I had to do all my homework by myself, whereas the cadets in the Queen Vic's would only have to walk down a corridor to find someone else to toss their problems about with.

At the opposite end of the scale was Mr Burrage, a jovial and good-natured individual who took us for ship construction and naval architecture. This subject could also have been exceedingly dull if taught by someone like Dr S, but during the course of a lecture Mr Burrage always managed to liven things up by throwing in an interesting anecdote or two about things that had happened to him during his seagoing career. A lot of this subject was devoted to calculations on ship stability, and involved some pretty complex maths. Ships are not a regular shape in any direction, and special formulae were employed (Simpson's Rule amongst others) in order to calculate water-plane and cross-sectional areas, from which all sorts of other data could be derived. This information could, for instance, be used to work out whether or not a new-build ship would float upright or capsize when launched, or what might happen to an existing ship if a particular compartment was flooded or a heavy weight lowered into one of the holds.

To make things a bit more interesting, the college had several large model ships floating in a water tank down in the basement, and we would occasionally be taken down there to put some of the theory into practice. The models were compartmentalised just like real ships, and these spaces could be wholly or partially flooded by opening little valve wheels mounted on the decks. On one occasion Mr Burrage had flooded a fairly large compartment up forward and to one side of a particular model, causing it to list about 15 degrees to port and also to go down by the bow. He asked me what I thought I could do to rectify this particular situation (and simply pumping the water out again was not an option, as the ship's hull was supposed to have been too seriously holed). My suggestion was to try a spot of counter-flooding on the opposite side in order to reduce the list. 'Off you go then,' said Mr Burrage, and I opened the appropriate valve. At first it seemed that I had done the right thing, as the vessel gradually returned nearly upright – but unfortunately it also sank further by the bow, until water came over the rail and flooded the forecastle as well, after which the ship gently sank bow first, in true *Titanic* style! We all had a good laugh at this while we waited for the model to be re-floated for a second attempt. The correct solution would have been to transfer as much fuel and cargo as possible back aft and counter-flood that end, which might have raised the bow sufficiently for the hole to be repaired.

One of the other topics we studied in naval architecture was the free surface effect. Free surface is what you have when a tank or compartment is only partially liquid-filled or flooded. If a ship rolls to one side, then the contents of any space will slop over to the same side and will increase the list still further (more, and very complicated, maths are required to work out exactly by how much.) This is what sank *Herald of Free Enterprise* in 1987, when the sea flooded into the vehicle deck through the open bow doors. The actual weight of water she took on board would not alone have been enough to sink her, but because of the large free surface all the water was able to rush over to one side, destroying her stability and causing the capsize. We were taught the maths to prove that by subdividing a tank longitudinally into two equal halves then the free surface effect would be reduced by 75 per cent. So, if a single longitudinal bulkhead had been installed down the length of the vehicle deck on *Herald*, the free surface effect would have been reduced by this amount and 193 lives would have been saved. At least in this case, then, the subject bore some relevance to what we might possibly encounter when we finally went off to sea, unlike the majority of the applied mechanics, thermodynamics theory and maths that we had to learn.

The maths and mechanics man was also a decent sort, despite realising that most of what he taught us would be forgotten as soon as we left college. He had been a stress engineer at an aircraft factory during the war, and once he recounted to us a job he had been given, which was to work out the maximum possible stresses on the engine mount-ings for a particular heavy bomber, and then from that to deduce the correct size of high-tensile bolts that were required to hold them on. The calculations took him nearly all day (everything was done with slide rules and log tables in those days) but he even-tually came up with a figure of 9/16" for the bolt diameters. Wearily taking the figures to the chief designer for approval, he was somewhat surprised when this worthy took one swift glance at them and said, 'That looks a bit small to me – better make them 5/8ths.'

This demonstrates an important engineering axiom – that to an experienced eye if something looks right then it probably is right, and the converse is also true. Take for instance two World War II fighter aircraft, both powered by the same Rolls Royce Merlin engine: one of those aircraft was the Boulton Paul Defiant, which looked clumsy and slow and indeed proved to be quite useless in combat – and the other was the Spitfire which, apart from being one of the most beautiful aircraft in the world, was the best fighter we had.

After a while, I got used to the college routine and had by now met another cadet, Peter Button, who also commuted from West Wickham, so I had company on the journey. If we managed to get away from college sharp at 5 pm and didn't have to wait more than a few minutes for all our bus and tube connections, there was a chance of getting an earlier train from New Cross, which would get us home by 6.30 pm – it was amazing how much better that felt. Of course it did mean a prompt getaway, and on the day when Dr S had us for the last period we seldom got it – I sometimes wondered if he had got wind of our travel arrangements and kept us an extra five minutes deliberately. Our dash for the train commenced with a sprint down a side street and onto the Commercial Road, where there were quite a few buses going in the right direction. Providing the traffic was reasonable we would be in Shadwell tube station by 5.15 pm. From Shadwell, the trains went either to New Cross, which we wanted, or New Cross Gate, which we didn't. If the first train to arrive was going to the Gate, we were faced with the dilemma of catching it anyway, which would mean an additional five-minute dash along the Lewisham Road to get back to New Cross, or waiting an indeterminate time for the next one. Either way, the extra time taken made the connection with our train at New Cross very marginal and we quite often found ourselves arriving on the platform red-faced and puffed out, just in time to see the back of it disappearing down the track.

We both had cars in which we occasionally commuted up to college, thereby giving ourselves an extra half-hour in bed in the mornings, although on the downside the Blackwall Tunnel, which we had to traverse in those days, was an extremely unpleasant experience. Before the second tunnel opened in 1967 the first was two-way traffic, and when heavy lorries met at the bends this could cause long delays in the suffocating exhaust fumes. My car was a 1937 Morris 8, which had cost me the princely sum of £10. It was very slow and not particularly reliable, so I seldom ventured up in it. Peter on the other hand, had a 1946 MG TC sports car, which was both dependable and great fun – how I coveted that car! After the first term, when our season tickets for the train journey had expired, we quite often came up to college in his car and split the cost of the petrol, which made it slightly cheaper than by public transport. Toward the end of the first year, Peter fancied something a bit more modern and faster, and bought himself a 1961 MGA. By dint of using up all my savings, selling the Morris and being given a small loan by my dad, I became the proud owner of Peter's TC for the sum of £130, which honed my driving skills no end and gave me a great deal of pleasure over the two years that I owned it.

The TC was not an easy car to steer accurately – the worm and peg steering box always had play in it, which meant that even on a straight road quite a bit of see-sawing at the wheel was required to compensate for bumps and changes in the camber. If you removed a shim or two from the steering box to take up the play in the centre, then the steering would become so stiff on bends as to make the car almost undriveable. For the annual MOT test, I would stop the car round the corner from the garage and take a shim out, so the car would not fail on slack steering, and then as soon as I had the test certificate safely in my pocket I would stop on the way home and put it back again. Happy days!

Lunchtimes at Poplar Tech were not exactly gourmet affairs. The college canteen provided egg and chips, pie and chips, sausage and chips, or burger and chips and (on Fridays only), fish and chips, all available with baked beans or processed peas at slightly extra cost. The pork sausages, I am sure, had never been anywhere near a pig, the burgers were like rubber and the pies were just gristle and gravy with a bullet-hard crust, so for me that just left the egg and chips – and even that could be quite disgusting, with soggy chips and the egg swimming in a puddle of fat. It seemed amazing to me that so many of my fellow cadets would happily put up with this muck every day of the week.

Luckily, Esso provided a few luncheon vouchers as well as the 7 shillings a day, and there was a cafe down the road that would take them. This was called the Nabi La La and was run by a miserable and somewhat malodorous Indian woman. The food was only slightly better than at college but there was also curry and rice on the menu, which was quite good and made a welcome change. The place had one other attraction for me, however (and I suspect for quite a few of the other cadets), and that was Mrs. Nabi's daughter, who used to do most of the waitressing. She was a really stunning Anglo-Indian girl and certainly the only thing of beauty I ever clapped eyes on during my two years in Poplar!

Once a week we spent the day in the college workshops, which were down in the basement, learning basic fitting and machining skills. We were each allotted a basic set of hand tools and a length of workbench with a vice, where we did the fitting work. In the machine shop there were enough lathes for each student to have one, but there were only a couple of milling machines and about four well-battered bench drills. Everything we made was marked and the figures totted up during the term, to give a final figure that would count as our OND exam result.

One of the fitting exercises was to make a set square from two pieces of steel accurately filed up and invisibly riveted together. Following assembly there would be a lot more filing and scraping to ensure accuracy, which we checked against a Moore & Wright master square. I was well pleased with mine, although I think it only scored a B minus and use it regularly to this day.

I still have the bell punch as well, from one of my lathe jobs. Basically this started as a hefty piece of round bar, on one end of which you machined a knurled handle and on the other a large hollow cone. This was then accurately drilled up the middle, and then an integral centre punch, which had to be made separately, was fitted to it. The idea was that you could place it over the end of a steel bar and give the punch a whack, and this would show you where the centre of the bar was. The only drawback was that the end of the bar had to be squared off in the first place to get an accurate mark – and how would you get it squared up? Well, the easiest way would be to put it in the lathe and face it off, which would find the centre for you anyway! Still, the bell punch was quite a pretty thing to make and I have this in my tool chest as well, although it has seldom been used.

Two incidents spring to mind from our workshop time: the first of these occurred when we were being shown a bit of forging. The exercise was to make a cold chisel (a chisel hard enough to cut cold steel) from a piece of hexagon section bar, one end of which we heated in the forge until it was bright red and then hammered it down on the anvil into the required chisel shape, after which it was hardened and tempered. One chap was busy bashing his red-hot lump of steel on the anvil but must have mishit it, for it sprang up in the air and then dropped down the back of his neck between his shirt and his boiler suit, which was fortunately rather on the large side. From there it slid right down his back, burnt straight through the seat of the boiler suit and dropped out. He wasn't even scorched, although his boiler suit, shirt and trousers all displayed evidence of his lucky escape.

The second incident was far more serious. It occurred in the machine shop, when one of the students was setting up a job in a four-jaw chuck on the lathe. With a four-jaw chuck, each jaw can be adjusted individually using a tee-handled steel chuck key, so that irregular shaped work can be held. The setting-up usually takes quite a few adjustments before the required position is obtained, and to make it easier the student was using the 'inch' button on the lathe to kick the heavy chuck round the quarter- or half-turn required between settings. This button would only move the lathe while it was actually being pressed, and as soon as you took your finger off, the machine would stop. On this occasion he was still holding the chuck key in the chuck with his left hand when he pressed the ''start' button by mistake. This of course made the lathe spring to life and the chuck and key whipped round together and squashed one of his fingers against the solid steel lathe bed. He screamed and jumped back, clutching his hand, before subsiding white-faced with shock on a nearby bench.

The last joint of his middle finger was completely flattened and split open. It was not bleeding very badly – no doubt on account of the blood vessels also being squashed – and it must have been numbed as well, because by the time the first aider had been summoned the victim had visibly brightened up and started to walk over to the sink to wash the injury. He met the first aider half way and exclaimed: 'Look what I've done, sir.' Sir took one look at the finger and promptly passed out, cutting his head on the corner of a machine as he fell, so the ambulance finished up taking them both off to hospital. Surprisingly, within a couple of months the finger had healed, although it was never quite the same shape again.

At the end of the first year, the college broke up for the summer holidays – unfortunately for us cadets, however, four weeks of this was supposed to be spent in further practical training, which was to take place at North Woolwich, in a workshop set up and administered by the college. The Woolwich Annexe, as it was called, consisted of an old factory building into which had been installed the most motley collection of marine engineering junk you were ever likely to see. Without exception it was obsolete and had been either donated by shipping companies when they had no further use for it, or been obtained from scrapyards. Hardly any of it actually worked, apart from a 6-cylinder Allen diesel generator engine, which stood about 8 feet high and made clouds of smoke and a most satisfying noise on the odd occasions it was fired up. We spent our time taking this lot to bits and putting it back together again without ever really feeling we were learning anything useful. The instructors, however, were a cheery bunch and, being several miles from Professor G's malevolent influence, didn't drive us very hard.

We spent another week of this summer holiday on the college training ship – SS *Glen Strathallan*. This vessel, which was operated jointly by Poplar Tech and the King Edward VII Nautical College (where the deck cadets went), had been designed as a deep-water trawler in 1928 but had never been completed in this guise as the builders had gone bankrupt. Eventually, she was bought by a millionaire and converted into a luxury steam yacht. When he died, he stipulated in his will that the vessel was to be used for training purposes, which is how she came into the possession of the colleges. She was only of 690 tons displacement and 150 feet long, so getting 25 cadets on board at a time was quite a squeeze, but we all managed to rub along quite amicably for the week. Most of our time on board was spent in the Millwall Dock, where the ship was normally moored, and our trip to sea consisted of a run down the Thames to Southend, where we anchored for the night before going back again next day.

I can remember very little of this week, except the fun and games we had trying to take indicator diagrams from the triple-expansion steam engine. An indicator diagram is a graph of the pressures in an engine cylinder at different points of the stroke, and from that graph the indicated horsepower can be obtained. Ours were produced by a cunning device called the Dobbie McInnes indicator, which could be connected to the cylinders at a tapping with a valve in it – an indicator cock. The fluctuating pressure in the cylinder caused a spring-loaded stylus to go up and down over a piece of paper wrapped round a small drum. The drum, in turn, could be rotated back and forth in time with the piston by connecting it by a piece of string to some part of the valve gear, where the travel matched the circumference of the drum.

The tricky bit was to get the loop on the end of the string hooked onto the peg on the valve gear, which would of course be going up and down in time with the engine at 130 rpm. Having done this, one had to get oneself into the rhythm of the engine and apply the stylus to the paper for just the one revolution. If you got it right the resulting diagram would be shaped roughly like a triangle with rounded corners and the bottom edge horizontal. By working out the area of the diagram, the mean effective pressure in the cylinder and hence the horsepower could be obtained. The engine was double-acting (the steam pushes on both sides of the piston in turn) so in order to calculate the overall power of the engine we had to take six diagrams in total, from the top and bottom side of each piston, and add the results together.

Some years later, the boilers on the ship needed major repairs and as the money could not be found the engine was removed and the vessel towed down to Plymouth, where she was scuttled and thereafter used for diving practice by the Royal Navy. The engine was installed in the Science Museum in London where it can be seen to this day – much to the amusement of my children ('Are you really that old, Dad?').

So ended the first year of my cadetship. The second carried on in very similar fashion, except that we were worked and tested even harder. The daily commute was still a real drag, especially as my friend Peter, who was a year ahead of me, had by now gone off to sea, while the Blackwall Tunnel traffic jam seemed to be getting worse than ever. I thought it might be a good idea to get a motorbike, which would both be cheaper on petrol and cut through the traffic, saving me some travelling time. I had never taken a motorbike driving test and my only previous experience had been on a

98cc New Hudson moped, which I had sold as soon as I had reached 17 and passed my car test. However, in those days you were allowed machines up to 250cc on L plates and I purchased a BSA C15 motorbike of that size, which was quite a bit heavier and faster than my moped. Although my brother was a fanatical motorcyclist, my heart was never really in it and I think that the old C15 must have sensed my lack of enthusiasm, because that wretched machine never once got me to college. Three times I set forth and three times it broke down on the way, the ensuing late arrivals at college causing me to drop still further in Professor G's estimation, if that were possible. I did eventually get it sorted, but a trip one October evening to see my brother near Tonbridge, when I got back home again with fingers and feet frozen and one trouser leg soaked with oil, persuaded me that enough was enough, so I sold it and have never been tempted to buy another.

For our final exams, we had a total of seven three-hour papers to sit – two each on Tuesday, Wednesday and Thursday, with the final one on the Friday morning, after which we were allowed the afternoon off – and we certainly needed it! The papers were maths, mechanics, electrotechnology, thermodynamics, naval architecture, workshop technology and technical drawing, while our practical work was judged on our course work, as previously mentioned. The results arrived quite quickly and I was pleased to see that despite Professor G's low opinion of me, I had passed in all subjects with a merit, except for the drawing, which was simply a pass.

Another four-week practical had been arranged for us for this summer holiday as well, only this time it was to be in Milford Haven. This was in a part of the country I had never seen before, so I was quite looking forward to the change. There were only about four of us reverse phase cadets doing this, because the rest were all going off to sea for their 18 months, so we fetched up being billeted together in a quite pleasant little guest house at Hakin, near Pembroke.

Milford Haven is really two ports: at the eastern end of the sound are the old Pembroke docks and harbour, while the huge oil terminal is a couple of miles further west. Our training was going to take place in the former. The dry dock and shipyard were not very busy, and the foreman, in whose charge we had been placed, was not particularly interested in us – rather than give us any specific tasks he more or less left us alone to wander round and look at whatever we fancied. With our ONDs safely under our belts, we were not inclined to work too hard, and if the weather was fine we would quite often manage to slip off early and find some quiet secluded beach where we could swim and sunbathe – all in all, it was the happiest time of my entire apprenticeship.

On the last day but one of the course, we had gone to Dale, which had a very pleasant and sheltered beach, just inside the mouth of the Haven. There was a particularly low spring tide that day, which had left quite a bit more of the beach exposed than we had seen before. Right at the water's edge I spotted a rusty old ship's anchor, which by the pattern, looked to have originated sometime back in the 19th century. This I thought would make a fine ornament to stand outside my parents' house in Shirley.

After promising to buy my three companions at least two drinks each that evening, I managed to persuade them to give me a hand to drag it all the way across the beach, up a steep path and back to the car, by which time we were all just about exhausted. The only problem now was getting it actually into the car, as it was so heavy that the four of us could only just lift it. By dint of removing the front passenger seat, we managed to heave it up so that the flukes spanned across the rear bodywork, sticking out slightly each side, while the shank went down into the passenger foot well, after which the spare seat was tied up on top in the back. In this condition the car was driven back to our guest house, where I left it parked outside as usual and prayed it wouldn't rain, as it was by now of course impossible to put the hood up.

Driving home the following day, I was stopped in Gloucester by a traffic cop riding a Triumph Thunderbird motorbike, who had no doubt spotted that my tax disc had expired the previous month. After getting off the bike and ceremoniously removing his helmet and gauntlets, he walked slowly right round the car before coming up to me, pointing to the anchor and saying (with a completely deadpan expression) 'What's this, then, sir – emergency brakes?' I couldn't help but laugh at this and after explaining that I had been away from home for four weeks and therefore unable to get the tax, he let me go. I never heard any more about it, which underlined his sense of humour. The anchor, by the way, has travelled with us each time we have moved house and currently resides outside our latest home, in Sussex.

After this pleasant rural interlude in Little England beyond Wales, as Pembrokeshire is sometimes known, it was back to the smoke, much to my disgust, as my year's workshop training was to take place back in the Woolwich Annexe again. The collection of marine junk was just the same as before, and I couldn't help but wonder what lies Poplar Tech must have spun to the shipping companies to convince them that we might learn anything useful there – which of course would mean the college getting another year's worth of fees.

This final year of my shoreside training was boring in the extreme and it didn't help when my friend Peter got back from his first voyage and regaled me with his tales of foreign ports, parties and girls. It all sounded wonderful, and I don't think that there were many of us who could foresee that these were the final few years of the traditional general cargo ship, and that within a decade the advent of containerisation would mean the end of the line for many long-established shipping companies – including Shaw, Savill & Albion, which was Peter's outfit.

To break the monotony we were occasionally taken to the nearby Royal Docks to visit a ship and see what an engine room actually looked like. One of these vessels had a five-cylinder Doxford main engine of a similar type to the model at college. To most people our Allen generator engine at Woolwich would have appeared to be a pretty big diesel, but if they had seen this Doxford, they would have been absolutely dumbstruck, for its height exceeded that of the average three-bedroom semi and it weighed well over 500 tons. One of the crankcase doors was opened for us so we could have a look at the connecting rod and big end of one unit – even the big end bolts were so big they looked as if they would be uncomfortably heavy to lift into place manually. Despite its huge size, the engine was not particularly powerful – about 5,500 hp, I believe – and its speed range was 25 to 120 rpm, which meant it could be directly coupled to the propeller without needing a gearbox or clutch. To go astern, the engine was simply stopped and then restarted in the opposite direction, using compressed air from a couple of huge air receivers (cylindrical tanks).

Eventually we got to the end of the final term and I went home for a few weeks' holiday, whilst awaiting the call from Esso to join my first ship. There was one last thing I needed, though, before I could go to sea, and that was to get my uniform. This meant a trip with my parents (who were going to have to pay for it) to Messrs. S.W. Silver & Co. in Fenchurch Street – not that far from British India's and several other shipping companies' offices. Esso had provided me with a recommended list of what they supposed I might need – and it was a pretty long list. First, there was the No. 1 uniform, consisting of the traditional doeskin jacket and trousers in dark navy blue (doeskin is a very finely woven woollen material with a nap rather like felt) and three long-sleeved white shirts to go with it. As it turned out, most of my time in Esso was spent going to and from the Persian Gulf (which was what it was called at the time – and I, like many others, still call it that) and other equally warm places, so I could count on the fingers of one hand the number of times I was to wear this on my first ship! Then there were three sets of tropical white shorts and shirts – and these I wore almost all the time.

There were quite a few other items as well, including three white boiler suits, a pair of stout engine room boots, a cable knit blue sweater and a white officer's cap. A lot of companies had their own cap badge, but Esso used the standard Merchant Navy badge with the crown, anchor and laurel leaves – I was quite pleased about that as an Esso badge on the cap, would I thought have made me look like a filling station attendant! The uniform was all of superb quality – I was still wearing the same number one uniform jacket at my daughter's wedding in 2014, while even the tropical shorts and shirts lasted over ten years, despite having been washed hundreds of times. In total it cost a small fortune and my poor father looked quite pale as he wrote out the cheque, but he never said a word, bless him.

Despite the final year kicking our heels at Woolwich, I think that in the main the training had been pretty good from both the theoretical and the practical aspects. When I eventually went aboard ship, I never felt in any way inferior to some of the other engineers I met who had done a five-year shipyard apprenticeship, which had been the usual form of training before cadetships had been introduced.

3 MY FIRSt SHIP

After my two weeks' holiday, the expected letter came from Esso, instructing me to report to Milford Haven on 14 August 1967 to join *Esso Yorkshire*. On the day before I was due to travel down, there was a phone call from Esso to say that due to problems with berth availability the ship would not be tying up until the early hours of the following day. I was now to travel just as far as Swansea instead, where a hotel room had been booked for me, and I would not join the ship until the morning of the 15th, when a coach would be provided to take me and a few of the other crew members down to Milford Haven. This meant I could take a later train from Paddington and have one last lunch with my mum and dad. Both my parents were very sorry to see me go, so lunch was a rather subdued affair, after which they drove me to East Croydon station for a fast train up to town. After saying our goodbyes, my mum thrust a gold St Christopher and chain into my hand, saying it was for good luck.

After the train trip from Paddington to Swansea (in which I took but little interest, as by now it was being hauled by a diesel locomotive), I took a taxi to the hotel, which was very plush, and I enjoyed a superb evening meal before turning in. This was something I found all through the years I spent working for oil companies: although they expected you to work hard on board ship, when you were ashore they would always look after you extremely well and no expense claim was ever challenged – at any rate, none of mine were.

The following morning, I boarded the bus with the other crew members and we headed west down to Milford Haven – I was somewhat embarrassed to find that I was the only one wearing uniform, and everyone else (including the captain, I found out later) was dressed in casual clothes. When we got our first glimpse of the haven, the ship could be clearly seen at the end of the jetty – even at a distance of at least two miles it looked enormous and completely dwarfed the various other vessels nearby.

At this time, the biggest oil tanker in the British Merchant Navy was *British Admiral*, which was owned by BP tankers and was about 100,000 deadweight tons. *Yorkshire* was not far behind, at 94,000. *Deadweight tons*, incidentally, which is always the figure quoted for oil tankers, refers to the carrying capacity of the ship rather than the actual weight. Dwt is the difference between the displacement, which is the vessel's actual weight when it is empty, and its weight when fully laden with cargo, fuel and stores. Passenger ships and cargo ships are, however, usually measured in *gross tons*, which is the volume of a ship below the upper continuous deck in cubic feet divided by 100 – it being reckoned that one ton of general cargo would take up 100 cubic feet of hold space on average. As a further complication, *net tonnage* is also sometimes quoted – this is gross tonnage less an allowance for the engine room and bunker spaces etc. which of course cannot be used for carrying cargo.

Close up, *Yorkshire* did little to inspire confidence, with the hull paintwork flaking off to give a patchwork quilt effect, and the deck and superstructure showing quite a few streaks of rust. It is supposed to be unwise to judge a book by its cover, however, and this was certainly true of *Yorkshire*, which proved to be a very well-found ship; as soon as I stepped inside the accommodation I found it all to be quite immaculate.

Having surmounted the gangway, I was directed down aft and told to report to the chief engineer in his cabin. The chief's cabin was in fact a whole suite of rooms, consisting of an office with a large desk, a huge day room with a three-piece suite, a bedroom with a double bed and a large fully tiled bathroom. The chief, who I think was Australian, said a few words of welcome after which he told me I would be put on the 4 to 8 watch and would take my orders from JH, the junior second engineer. Apart from one occasion, when I was hauled up before him for a certain misdemeanour (more of this later), these were the only words he spoke to me during the entire three months I was aboard that ship!

Following this very brief introduction, I had to go down to the mess room to sign 'articles' for the first time. Ships' articles are legally binding documents that basically put you under the rule of the captain for the duration of the voyage, or until the ship returns to a UK port (where you could ask to be relieved), or until the articles themselves expired. They could not be extended indefinitely, as otherwise a ship owner could ensure the ship never came back to the UK and the seamen concerned might be stuck away from home for years. Your date of signing was recorded in your discharge book, which was a blue book very similar to a passport and was countersigned by the captain or the shipping agent and stamped with the ship's official stamp. When you signed off again (discharged from service, hence the name of the book), you got another stamp. If you were unhappy or had any other reason for leaving, you could not simply get off at the first port of call, because you would not get the necessary stamp in your discharge book (this was called jumping ship, and could mean you would never get another job at sea). The book itself was kept in the captain's safe all the time you were on board, and was only returned to you when you signed off.

After this ritual engagement process had been completed, I started asking round where my cabin was situated, only to be told that because all the cabins down aft were already occupied, I was to be billeted amidships in a spare navigating officer's cabin. This too was very comfortable, with a writing desk and chair, an upholstered three-seater bench attached to one bulkhead, and on the opposite side a comfortable bunk. It also had an en suite toilet and shower room – in fact everyone on board, right down to the galley boy, had an en suite.

Yorkshire had been built in the old-fashioned style with the bridge superstructure, including the captain and deck officers' living quarters, placed amidships, whilst everyone else resided down aft in a separate accommodation block about 130 yards away. They were connected by a walkway raised about 8 feet above main deck level, called the 'flying bridge'. This provided the only safe transit between the two in heavy weather when seas were sweeping the deck. There were two steel shelters, called 'bus stops', evenly spaced out on the flying bridge, so that anyone walking down it who spotted a big wave coming would have somewhere to take cover, although they could still be in for a soaking because the shelters were open at both ends. Even a trip down the flying bridge could be quite hazardous if the sea was really rough: a few years later I was on a ship where one of the bus stops had been bent over at 30 degrees by a wave! To be down on the main deck in such conditions would be suicidal.

Having unpacked my kit, I saw it was time for lunch, so I trotted back down the flying bridge to the officers' mess, which was quite a spacious room, in Scandinavian style (the ship had been built in Sweden) with light-coloured wood and a minimum of decoration apart from a framed portrait of HM the Queen on one bulkhead (every ship in the British Merchant Navy has one of these). There were about eight tables, all laid up with spotless white tablecloths and silver cutlery. The steward, wearing a smart white uniform with brass buttons, presented me with a menu and waited to take my order. This took a little while because there was a lot to choose from: soup, choice of about three main courses which included curry and rice (this was always available at lunchtimes), various salads, and a hot dessert or ice cream, with cheese and biscuits and coffee to follow – if you had any room left! On nearly all the Esso ships I served on, I found the catering first class, and the hardest choice to make at meal times was usually what not to have.

After a splendid lunch, I thought I had better put on a boiler suit and go and have a look at the engine room, although I wasn't actually required to be on watch until 4 pm. There was a main corridor, which went nearly all around the aft accommodation block, from which you could get into the engineers' changing room and thence to the engine room. In common with most of the other engineers, however, I preferred to get changed in my cabin, and only used the changing room to store my dirty engine room boots, which were of course prohibited from the smart accommodation. From the changing room, a heavy steel door opened into the engine room itself. I will never forget the moment I opened this door and stepped through it for the first time – it was as if I had somehow got lost in a smart hotel and upon opening an unfamiliar door, had found myself at the top of a huge underground power station – which apart from the fact we were on a ship, is what it really was.

I had entered the engine room at one level below the very top, and found myself on a wide steel deck which went right round the inside of the accommodation block at this level, with guard rails protecting a large opening in the middle. It was rather like some modern buildings with an atrium in the centre going from the ground floor right up to the roof, with all the working floors off to the sides. Above my head were the engine room skylights, which consisted of some four pairs of hinged and louvred flaps that could be lifted pneumatically to let some of the hot air out – which it certainly needed to, because the temperature at this level was around 90°F (32°C). Below the skylights, it must have been a clear 60-foot drop to the main turbines and control console, in front of which a couple of tiny boiler-suited figures could be seen standing at a desk and drinking tea.

From my deck there was a lift down to the control platform level but I opted to use the ladders instead. Engine room ladders are not like stairs in houses: for a start they are much steeper, normally having an angle of at least 60 degrees to the horizontal instead of the usual 45, and secondly, the treads themselves are an inch or two further apart. It was seven flights and 108 steps down from the changing room to the lowest level in the engine room. Although it had been drilled into us at college that the only safe way to go down a ladder was backwards and facing the treads while keeping a hold on the handrails above your head, I soon found that nobody ever did this. Unless you were carrying something heavy or awkward it was much quicker to go down facing forwards with just one hand behind you to hang on with. Quicker still (and quite good fun besides) was to simply face forwards, grab the polished steel handrails with both hands and slide down with your feet held out in front and your arms locked straight beside you. With practice I could get down the ladders to the control platform in about 15 seconds, which was quite a bit quicker than the lift!

Having arrived at the control platform, I introduced myself to the two figures I had seen from above, who turned out to be the third and fifth engineers on the 12 to 4 watch. Quite a few of the older shipping companies just kept going with their rankings, so you could sometimes find a ship with seventh, eighth and even ninth engineers, which I thought was just plain silly. In Esso, however, any engineer below fourth was a junior or fifth engineer (often called simply a 'fiver') and *Yorkshire* had three of them, one on each watch. The pair I met were both Geordies, which turned out to be quite common in Esso, and for the first few weeks I had a bit of a language problem, which was not helped by the fairly high level of background noise in the engine room. They were very friendly, though, and made me a cup of tea as well, before the junior gave me a quick tour of the engine room, which I will describe in some detail here, so you will have an idea of what I will write about later.

Although we were still in port, there was quite a lot of machinery running – a 2 MW turbo alternator and three 2,000 hp turbine-powered cargo pumps for starters, so both main boilers were fired up and delivering steam at the incredible pressure of 750 psi (about three times more than the highest-pressure steam locomotives). Although this was a modern ship (modern for the sixties, that is) there was very little in the way of automation: the control console was like a great stainless steel wall about 20 feet wide and 7 high, completely covered in dials and gauges which had to be continuously monitored so that any irregularities in the readings would be quickly spotted and the appropriate response made. This was why there always had to be two men on watch, so that if one of them needed to go somewhere to make an adjustment or check something out, the other could remain at the control console keeping an eye on things – or 'minding the shop' as we called it.

In the middle of the control console were two large chrome-plated steel handwheels, each something over 2 feet in diameter – these were the main engine throttles, one for ahead and one for astern. Just beside the throttles there was a traditional engine room telegraph instrument that transmitted the bridge requirements to the men on watch –

Slow Ahead, Stop, Full Astern etc. Over to one side of the console was a wooden desk upon which rested the engine room log book; it was quite a big document with each page about twice A4 size. Every day, starting at 12 noon, a fresh double page would be commenced. In the book were recorded scores of pressures, temperatures, tank levels, the boiler water test results and even an analysis of the flue gas. Although some things were only recorded once a day by one particular watch – the boiler water and flue gas tests for example, being the responsibility of the 12 to 4 – there were still an awful lot of figures that had to go in every watch, so I could see that quite some time would be have to be devoted to filling it in.

Some way behind and below this position a few feet of the polished steel propeller shaft could be seen, from where it left the reduction gearbox down to a bulkhead, where it disappeared into the shaft tunnel compartment. It had a series of white marks painted on it in one place, making it look rather like a wrap-round zebra crossing, so that the engineer at the wheels could actually see whether it was turning or not – and it was now, albeit very slowly, being driven by the electric turning gear motor while we were in port. In fact, apart from the occasions when the turning gear itself was actually being engaged or disengaged, which only took about 30 seconds, the shaft never stopped at all during the entire time I spent on board. This was because the clearances between the fixed and moving parts of the turbines were extremely fine and if the rotor was allowed to stop before it had completely cooled down (which would take a lot longer than the average time a tanker stayed in port), the resulting uneven cooling could cause enough distortion for an unwanted – and expensive – contact between the two.

Considering that the turbine engines developed some 24,000 horsepower, they were remarkably compact, the high-pressure turbine casing being only about 6 feet long and 4 feet in diameter, including the lagging. Steam from the boiler (at 750 psi, superheated to 500°C) went via the main throttle to the front of this turbine and then through a whole series of nozzles and blades inside the casing to the aft end, by which time some 12,000 hp had been developed. Having done its work there, the steam was very much reduced in pressure and increased in volume and it exited via a large, heavily lagged pipe, which did a great U-bend up, across and back down to the LP turbine alongside it, which was about 4 feet longer than the high-pressure one and twice as fat. Here it went through more nozzles and blades, whose size steadily increased as the pressure dropped still further, and in doing so produced the other 12,000 hp. When the steam exited the final set of blades, the original 750 psi had become a vacuum – typically 27 to 28 inches of mercury. The exhaust steam from the LP turbine went straight into the condenser, positioned immediately below both turbines, where it was turned back into distilled water. Finally, via a long and complicated feed water heating system, this water was pumped back into the boilers again.

The LP turbine was actually two turbines on the same shaft – the ahead section, as described above, took up most it, but when the vessel was required to go astern, steam was supplied from the astern throttle valve to a separate set of nozzles and blades on the other end of the rotor. The astern turbine, used when entering or leaving port, had only had a few rows of blades and was extremely inefficient; to achieve a full astern movement of 60 rpm at the shaft needed even more steam than when running full ahead, which under manoeuvring conditions was 75 rpm.

Although the turbines themselves were extremely small compared with a diesel engine of even a quarter their power, all the other pieces of machinery that comprised the plant as a whole more than made up for this. The two boilers, for a start, stood around 50 feet high from the water drums at the bottom to where the uptakes (exhaust pipes) emerged at the top. The main condenser had at least the same volume as a double-decker bus, and behind that was a huge gearbox which reduced the 6,000 and 3,500 rpm of the HP and LP turbines respectively down to the much lower revs required by the shaft: at normal cruising speed this was 105 rpm.

From the control platform another ladder led down to the lower levels of the engine room – the bottom plates as they were known, which was where a great deal of the auxiliary equipment was mounted, including the biggest electric pump I had ever seen: this was the main circulating pump, which provided the seawater cooling for the main condenser. The bit you could see, with the 400 hp motor on top, stood at least 8 feet above the plates. A vertical shaft emerged from the bottom of the motor and went down to the pump end, most of which was below floor plate level, so that in total the machine must have been about 14 feet high.

The floor plates themselves were made of steel into which a diamond pattern had been pressed to give a better grip for the feet – this was called chequer plating, and was a feature of every engine room I ever went into. The plating was screwed down onto angle iron frames which were raised in most places about 5 feet above what we called the tank top (the ship was double-bottomed throughout the machinery spaces, and the tank top was in fact the inner bottom plating). Beneath this were tanks for fresh water and lubricating oil, or permanently empty compartments which were known as void spaces. This area, below the floor plates, was of course as low as you could actually get in the ship, and will be known to most people as the bilges – it was an area with which I was to become intimately (and uncomfortably) acquainted over the next few months, as it was stuffed full of pipes and I was supposed to find out what they all were and where they went.

At the forward end of this area another few steps led down to the cargo pump flat, where the steam turbines driving the three cargo pumps and the one ballast pump could be found. The output shafts from these passed through a bulkhead into the pump room where the pumps themselves were situated. I was told that each of the cargo pumps could discharge nearly 2,000 tons of crude oil per hour when working flat out. Because of the risk of fire or explosion, there was no direct access from the engine room into the pump room, and so to get there you had to go up on deck and then down again on the other side of the bulkhead. The pumps were the responsibility of the deck department, and unless there was a problem with them, the engineer officers seldom had cause to visit the pump room.

At the other end of the bottom plates on the starboard side, a power-operated watertight door led through into the shaft tunnel compartment. On general cargo ships, where the engines were situated nearer the middle of the ship, the shaft tunnel really was a tunnel, possibly a third as long as the ship herself, with a walkway down one side and just enough room to stand upright. On tankers, however, where the engines are right down aft away from the cargo tanks, there was no tunnel as such and the shaft compartment could be quite a large space: on *Yorkshire* it was about 70 feet wide at the forward end, tapering down to about 10 feet aft, and was also some two decks high. The massive steel shaft was highly polished and this finish had been achieved in a rather domestic fashion, by hanging a couple of coconut door mats across it, secured in place by ropes tied to the guardrails alongside. Once a watch, the duty rating would shift them along a bit, maintaining an even shine.

Climbing back up to the control platform level, I was taken down to the port side, where the two turbo alternators and main switchboard were situated; these provided and distributed all the ship's power. The alternators produced 3-phase current at 440V for all the main motors on board the ship, but it was transformed down to 110V for the accommodation lighting, power points and galley equipment. The alternator in use was spinning round at 3,000 rpm, and the turbine driving it at several times that speed, but the set as a whole ran so smoothly that a florin (a pre-decimalisation coin equivalent to 10p in today's money) could be made to balance upright on the bedplate.

Next we climbed back up again to visit the boiler room, which was situated behind and above the turbines and was reached by short ladder leading up to a door in the forward boiler room bulkhead. *Yorkshire* had two water-tube boilers of Foster Wheeler pattern, each with four burners; at full speed they consumed some 140 tons of heavy fuel oil a day between them. Each of the burners had a spy hole with a mauve glass window through which the flame could be observed – at present each boiler had two burners in use. Some 12 feet over our heads, the ends of the steam drums could be seen, each with two water gauges of the Klinger pattern, having square-section glass mounted in a heavy steel frame. Beyond that there was at least another 40 feet visible before the uptakes disappeared through the deckhead and into the funnel.

The boiler room had another stainless steel console covered in gauges. It was quite a bit smaller than the one in the engine room, and did not require to be supervised continuously except when the ship was manoeuvring, at which time two ratings were required to operate the burners. How it worked was that the engineer officer at the engine room console would assess the steam demand and operate a row of switches, each of which illuminated a light on the boiler room panel to let the ratings know which burners to fire up.

One interesting feature of the control console here was that there were two periscopes, which extended right up to the top of the boiler room, where they turned through a right angle to let you take a squint into the uptakes; they were lined up with light bulbs in housings on the opposite side. If the boiler was making smoke, which always happened when a burner was first ignited, the bulb would be obscured and the engineer's view at the platform below would suddenly turn black, whereupon he (in those days it was always a he) would adjust the combustion air until the periscopes were clear again.

My conductor led me steadily upward, until we were at the very top level, where the periscopes finished up. There was a thermometer hanging on the handrail here, which I noted was reading 110°F (43°C). This was obviously the place where the engine room department chose to dry their washing because all the handrails were festooned with boiler suits and other gear. Probably because the humidity was very low and there were ventilator outlets conveniently placed to blow some cooler air across the space, it did not feel too unpleasant, although I was still glad when we exited through a side door and got out on deck. I was to find many times over the years that it was not so much the actual temperatures that could make the working conditions unbearable as the humidity and whether or not there was any ventilation.

After taking a little turn around the aft deck to get some fresh air and have a look at the mooring winches, we started to go down again, this time via the steering flat. The steering system consisted of a pair of huge tiller arms connected to the top of the rudder shaft, the arms operated by four hydraulic rams working in opposition and powered by a pair of variable-delivery hydraulic pumps. At sea just one pump and one pair of rams were used, but for manoeuvring both sets were employed, to give a quicker response to the helm. I also noted a ship's wheel and a gyro compass repeater up by the forward bulkhead, which could be used in emergency to operate the steering system directly from the steering flat, if for some reason the main controls on the bridge were out of action.

From here, we made our way back to the control platform, paying a brief visit to the workshop on the way. This, I thought, was very well equipped, with a substantial lathe, drilling machine, hydraulic pipe bender and both gas and electric welding equipment. The upper half of the bulkhead on two sides had been made of heavy gauge steel mesh, so that when you were working inside you could see out over the engine room and control platform. The latter, incidentally, was hardly ever referred to by this name but was simply called 'the plates', and the watch keeper in charge was said to be on the plates or walking the plates.

This, then, was to be my place of work for the next few months, and to say that I was somewhat overawed by the complexity and sheer scale of it all would be an understatement, especially as there was still quite bit of it I had yet to see, such as the air conditioning and refrigeration plant and the emergency diesel generator room, not to mention the rest of the ship forward of the engine room. How anyone could possibly learn how it all worked and where all the pipes went was, I felt, quite beyond me at that time.

I didn't have too long to worry about all this, however, because a stocky, blond-haired figure could be seen descending the ladders toward us, who, the third engineer informed me, was JH, the junior second engineer and my immediate boss. After a few words of introduction and establishing that this was my first trip, JH told me that for this first watch I was just to follow the junior around and see what he did, after which I would be expected to start learning all the ship's systems in earnest until I knew the place inside out – or at least as well as *him*, he said somewhat disparagingly, indicating the retreating figure of the third engineer, who had now finished his watch. I was to find out that he appeared to have the same superior attitude to anyone he regarded as less intelligent than himself – which meant just about the entire ship's company – and that he was a very hard man to please. He was also quite arrogant, and would tell anyone who was prepared to listen that he had been the top cadet in his year at college and was the youngest holder of a chief's ticket (Dept. of Trade First Class Engineer) in Esso, if not in the entire Merchant Navy.

The first thing we did was to have more tea, after which I trotted off behind the junior (I am afraid I have long since forgotten his name but I will call him Fred from now on) for another inspection of the plant – following a similar route to my previous trip but looking at things in greater detail – I noticed for instance that Fred put his hand on various bearings, to check their temperature as he went round. Going round the job like this was invariably performed shortly after taking over a watch, and would take between half and three-quarters of an hour. As we would be out of sight of the control panel for quite a lot of that time, the engineer minding the shop would, if he needed assistance, summon us back again by setting off a klaxon alarm that sounded throughout the machinery spaces.

Fred was another Geordie and had entered the profession after a five-year apprenticeship in a shipyard on the Tyne: because of this he had not received any of the theory training that an engineer cadet would have had and was therefore yet another person that JH didn't rate too highly – despite, as I was to find out later, being a superb welder and machinist.

This second trip around our complicated domain and one or two other jobs, such as transferring some fuel and running up the ballast pump, took us halfway through the watch, which meant it was time for dinner. This was served from 6 to 7 pm and the normal arrangement was for the 8 to 12 watch keepers to relieve us for half an hour so we could get our meal and then they would go up and have theirs. Accordingly, just before the hour, the fourth engineer and his junior appeared from the lift and after a few words to let them know what was going on we went up to get changed; in my case this meant another trip down the flying bridge to my cabin and back again.

By now, I had found that apart from on special occasions such as Christmas Day, or when we had visitors from ashore, nobody wore full uniform in the mess and that the navy blue trousers and a white shirt were sufficient. The captain, I noted, was one of the most casually dressed of all and appeared to be in the same check shirt as the one he had been wearing on the coach down from Swansea.

I was also surprised to see three women sitting down with us at dinner: it turned out that these were the wives of the second mate and third engineer, who would be coming along for the voyage, and the chief officer's wife, who was just visiting while we were in port. It was company policy for all the officers to take their wives with them at any time they liked, and their cabins all had double berths.

Dinner was another splendid meal, the main difference from lunch being that there was now a fish course as well, and that, as I was to find out, one of the main courses was always a roast. I managed to dispose of the soup, the roast – pork, I seem to recall – and then some ice cream, before it was time to whizz forward to get into my boiler suit again.

The second half of the watch was largely spent following Fred around again while he collected all the data required for the log. He carried a piece of card – I think it might have been an empty fag packet opened out – upon which he had written his own particular brand of shorthand to indicate which reading was which. After returning to the plates, he copied the figures into the log. JH told me to take careful note of it all, because when we got to sea it was going to be my job. I didn't much like the sound of this, because I could see that it was all done in black ballpoint and that any mistakes made would be rather obvious.

As the time approached 8 pm, the fourth engineer and his mate appeared again to relieve us, and we all went topsides in the lift, after which I traipsed down the flying bridge yet again to my cabin and a welcome shower. A little while later, there was a knock at the door, which was Fred, to say that a few of the lads were going ashore for a drink and did I want to come? Accepting with alacrity, I joined the others on the jetty, where a taxi had been ordered to take us into town. I can't remember much about the evening that followed (not because I got plastered, I hasten to add, but simply because it was half a century ago) except that I managed to trip over a steel mooring cable on deck when we returned and tore the new trousers. So ended my first day on board ship. The following morning we would sail for the Persian Gulf, stopping at Las Palmas in the Canary Islands en route to take bunkers.

4 to SEA At LASt

I was jolted awake at 3.30 am by one of the seamen, who stuck his head round the door, yelled, 'Time to go on watch!' and turned the light on to make sure I didn't nod off again. After this shock to the system, I crawled blearily out of my bunk and got dressed, which didn't take very long because all we ever wore in the engine room was underpants, socks and a boiler suit – it was always warm enough down below to preclude the need for anything else. The transit of the flying bridge at this hour when so lightly attired was a bit shivery, however, and I was glad to step into the warmth of the accommodation at the end of it.

In case the superb meals that had been provided for us were not enough, there was also a pantry alongside the mess, which was stocked with bread, cheese, cold meat etc, in case anyone felt a bit peckish in the night and fancied a bit of toast or a sandwich. Fred was already in there and had the toaster fired up, so I joined him and stuck a bit on for myself as well. The ship's cooks baked fresh bread for us every day, and I usually found that the first job before making toast was to carve a wedge off the end of the loaf to straighten it up – it never ceases to amaze me what a mess some people can make of cutting a slice of bread.

When we arrived down on the plates we were informed by the third engineer that there had been a problem with one of the stripping pumps and that our sailing had now been postponed until the next high tide. I should explain here that the main cargo pumps which were used to get the bulk of the cargo ashore were centrifugal, so as soon as the tank level had dropped enough for any air to be drawn into the inlet pipes they would lose suction and stop pumping. When this occurred, the main pump would be stopped or the suction changed over to another full tank, and a stripping pump started in its place. The stripping pumps were vertical steam reciprocating pumps of the Weirs type, built to a design going well back into the 19th century. These, despite their antiquity, were usually very dependable and they would pump a cylinder full of whatever came in, including air, at each stroke, so they were ideal for getting the dregs out of the tanks.

I seem to remember that *Yorkshire* had three of these pumps, and as these were now reduced to two this was why we would miss the tide. There was nothing we could do from the engine room side to get the stripping pump started again, nor would we be allowed to enter the pump room to effect repairs until it had been gas-freed, which would be several days into the voyage. This was in case we dropped a hammer or spanner down the 70-odd-foot-deep pump room, caused a spark and blew ourselves to kingdom come.

Apart from the bigwigs in head office who were always trying to reduce our already brief stays in port, nobody on board was too concerned at this development – especially me, as it meant I would have a bit longer to learn my way round the ship before we put to sea.

Out of the three watches, the 4 to 8 was without doubt the best one to be on providing you could get used to the early start, and it was usually taken by the senior watch keepers – in this case, the junior second for the engine room and the first mate on the bridge. We had a senior second engineer as well, but he was on day work and apart from stand-bys, when he would assist the watch keepers, he basically worked 8 am to 5 pm. In fact I hardly ever saw him in the engine room at all, and he seemed to spend most of his time in the engineer's office doing paperwork – probably the chief's paperwork if truth be known. Doing the 4 to 8 meant you could spend a few hours enjoying any social activity that might be going on in the evenings and still get a decent sleep before the morning wake-up. If this wasn't enough, you could always take a nap after lunch and get another couple of hours in your pit, as your bunk was usually called.

Of the other two watches, both had their disadvantages: the 8 to 12 was very antisocial because any party in the bar would usually be over by the time you had come off watch and got showered, so unless the second mate from the corresponding bridge watch decided to come down and join you, the engine room watch keepers would find themselves drinking alone in the bar, some time after midnight. You also had to have what we called a seven-bell breakfast (half an hour before it was served for everyone else at 8 am) which, apart from the second mate once again, the two of you would eat in solitary splendour in the mess. On the credit side you could usually get a solid six hours sleep in one hit, which you wouldn't on the other watches.

It was the 12 to 4 however, that I really hated because then I never seemed to be able to get enough sleep. I have always found it difficult to nod off if I go to bed early, which is what you had to do on this watch – the captain and chief engineer could usually be relied upon to give their respective watch keepers a disapproving look if they were still in the bar much after 8 pm. If I turned in at this time, however, it would always take a couple of hours before I actually got off to sleep, only to be woken again an hour or so later to go on watch, feeling worse than if I hadn't been to bed at all. In the mornings it would be a case of having a shower and possibly a couple of beers in the bar, with the aim of being in bed by 5 am – unless you thought that a few more beers might help you sleep, in which case you could stay up and watch the sunrise, which I have done on many occasions. Either way you wouldn't get a lot of sleep because at 7 am the ship started to come to life and there would then be a series of doors banging, footsteps down the passageway and possibly even the steward with his vacuum cleaner to wake you up. If you were really unlucky, the deck department might decide that the decks needed a bit of chipping and painting, and the racket from the pneumatic needle guns and hand chipping hammers when they started up after breakfast was enough to wake Rip Van Winkle.

This time, however, the morning watch passed quickly enough, with me once again trotting along behind Fred, like a little dog. One place we visited that I had not seen before was the fridge and air conditioning flat, which was situated forward and to starboard of the steering gear. Adjoining the actual machinery there were several large walk-in freezer and chiller rooms, which could store enough provisions for several months. The temperatures of these rooms were always recorded in the log on our watch, and although they were padlocked, the engineers were provided with a key to check inside in case any of the fridge equipment developed a fault in the night – and also to purloin a tin of pineapple chunks or some similar treat if we felt like it, Fred told me with a wink.

The chief steward would take every opportunity to replenish fresh fruit and vegetables, and the only thing on board ship I really missed in the way of food and drink was fresh milk. This was before long life milk was available in the UK, so when the fresh stuff ran out a few days after leaving port, we would be reduced to condensed or evaporated milk from tins, both of which I thought were disgusting in tea or coffee, although cocoa made with condensed milk wasn't bad. We also had milk powder, which was mixed up with water in the galley every day using a machine which we called the 'iron cow' – this could supposedly be used in the same way as fresh milk. I was none too keen on this, either, especially on the breakfast cereal, although it was a half-decent whitener for tea and coffee.

After our watch, JH told me to have breakfast and then report to the engineers' office, where he would outline in more detail what was expected of me. The breakfast menu once again was amazing: you could have a fruit compote (stewed apricots, prunes and figs), a choice of every cereal known to Kellogg's, fruit juices, porridge, kippers or smoked haddock, full English with as many eggs as you wanted cooked any way you wanted, toast or fresh baked rolls, various jams, marmalade or honey, and tea or coffee. I was quite hungry after the watch and managed all of it apart from the fruit compote and porridge. This impressed the steward no end – 'Right little gannet you are,' he said in his gentle Welsh accent.

He then asked whether or not I wanted anything from the bond when it opened after we had put to sea. I was about to order a half dozen tins of beer until he showed me the price list, when I found out that a case of 24 cans of Tennent's lager, which was the only sort they stocked, was a meagre 18 shillings (90p in today's money), so I ordered a case instead. The cans were quite a bit smaller than the average supermarket can today – 340 ml as against 440 – but even so, this was amazingly cheap. A bottle of whisky was 10 shillings, while rum and gin were only 8 shillings! It was little wonder, then, that heavy drinking on board ship was not an unknown occurrence.

After breakfast I met JH as requested in the engineers' office, where he gave me a foolscap-size hardback notebook in which I was to record my line drawings of all the ship's systems as I completed them – but not before he had personally checked and approved them first. I still have this book, which never fails to bring back unhappy memories of the time I spent producing it.

After this, he said I was to assist Fred in the workshop, where he was repairing a large gate valve from the bilge system. I had fondly imagined that I would now be off duty until 4 pm, so this came as a rather unpleasant surprise. Our normal watchkeeping routine of four hours on and eight off gave a 56-hour week anyway, so I couldn't see why we should be expected to do any more except in an emergency. Having to do extra hours after a watch was called working a field day, and it seemed that on *Yorkshire* every day was a field day except Sunday, when JH relented.

Down below once again, I found Fred working on the lathe, turning up a new nut for the spindle of this gate valve because of a stripped thread on the old one. This meant screw-cutting an internal Acme thread in a 3-inch diameter lump of bronze that he had hacksawed off from a stock length. My contribution to the project was simply to cut out new joints (never called gaskets on a ship) and clean up all the threads on the nuts and bolts that went with it. Despite having to grind up a suitable lathe tool for the job before he even started, he had the nut finished by 10.30, which I thought was damn good going and made him well worth his place on the crew despite JH's low opinion of him. Another hour saw it reassembled and back in place by the bilge pump, and then we really were finished for the morning. Fred invited me back to his cabin for a beer before lunch, which was good of him, as I was to find in future that quite a lot of the older engineers who had served their time in a shipyard didn't think very much of engineer cadets, and treated them simply as cheap labour. Fred was quite chatty, however, and asked what had made me decide on a life in the 'treacle mine', which was how he described the engine room. I had never heard it called this before, but found that it was quite a common expression amongst the Geordies, who made up at least half the engine department. Seeing as most deep sea ships at this time were burning heavy fuel oil, which looks exactly like black treacle (and is just as thick when cold), perhaps it is not a bad description – and it gave me the title for this book.

We chatted a bit about our respective backgrounds, and I learnt that his grandfather had spent all his life working in the Tyne shipyards as a riveter and was now almost completely deaf as a result. My mentor was at pains to point out to me that it would be quite easy for us to go the same way if we didn't take steps to protect our hearing – to show he was serious, he usually wore cotton wool plugs in his ears when on watch. At that time very little research had been done into hearing loss due to noise exposure, and it was to be quite a few years before I went on a ship where proper ear defenders were provided for the engineers.

After lunch I took a stroll down the jetty to find a phone box and let my Mum and Dad know I had settled in – they were impressed when I told them about the food. With mobile phones still nearly 30 years away, it would be the last time I would be able to speak to them until we got back to England again, and I had quite a lump in my throat when I finally hung up and wandered back to the ship.

Back down below again at 4 pm, the third told us that we were expecting to sail at 1845 hrs. 'Can't have the Old Man and the chief missing their dinner,' was his parting comment before going off. After the usual walk round the job JH said it was time to get a few things ready, which included raising vacuum in the main condenser and putting on the second alternator, which took us pretty well up to dinner time. Something else I learnt then was that anything that produced electricity on a ship was always called a jenny (short for generator), although strictly speaking the term 'generator' should only apply to a machine that makes direct current (DC) as opposed to our alternators, which produced alternating current (AC). Even back in 1967, ships running on DC were very much in the minority.

Returning after our meal, it was time to start warming through the main engines. First off we phoned the bridge to obtain permission to turn the propeller, in case there was anything fouling it or likely to be upset by the wash. Next, we stopped the turning gear motor and disengaged it; this was done by removing a padlock from a lever, moving the lever over to the out position and then padlocking it up again – this to preclude any possibility of an almighty and expensive grinding of gears if it was put in by mistake when the engines were running!

Finally, we opened the bulkhead stop valve, which allowed superheated steam through from the boiler room to the throttles. Then JH took the ahead wheel and spun it open a couple of turns until it met some resistance, after which another, gentler, half-turn produced a muffled hissing sound and the tachometer wobbled off the stop and crept up to about 10 rpm – a look back at the zebra stripes confirmed the shaft to be gently rotating, although the turbines themselves made virtually no sound to indicate they were running. After a few seconds, he shut the ahead throttle and opened the astern one instead, until the shaft started turning the other way. After doing this a few times he handed them over to me, with the instructions not to exceed 10 rpm and never to allow the rotors to remain stationary for more than a minute or so at a time. This little chore kept me employed for the next 20 minutes, which brought us up to our predicted departure time.

Promptly at 1845 hrs, there was a phone call from the bridge to ask if we were ready, and upon being told that we were, the telegraph clanged round from Finished with Engines to Stand By. Nothing happened for 5 minutes, then JH said, 'Won't be long now, they're heaving the ropes in.' When I asked how he could possibly know this, he indicated the rows of gauges on the console and said, 'It's there for all to see if you know where to look.' When I still looked puzzled he indicated the deck steam pressure gauge, which was reading 115 psi: 'That's 5 pounds below normal, which means they must be using steam for the winches.' That he could pick out a drop of 5 psi on one gauge out of the entire panel was, I thought at the time, quite impressive.

We had been joined by the electrician and the extra third for stand-by, the latter taking charge in the boiler room, while JH left me on the wheels – boy, did I feel important, with 24,000 hp in my hands! JH's prediction proved to be accurate, as very shortly afterwards the telegraph rang Slow Astern, which I answered before starting to wind open the astern throttle. JH enjoined me to watch the steam pressure gauges for the turbine inlets when manoeuvring rather than the tachometer – 'It's steam that drives these engines, not that bloody rev counter' was the comment – and I found that someone had put tiny red paint marks on the relevant gauge rims, which corresponded pretty well with the desired rpm.

There were a couple of Slow Astern then Stop movements, all of which were recorded in the movement book, after which the telegraph rang for Half Ahead, followed shortly by Full Ahead. JH said, 'That's it; we're away now until they drop the pilot,' and enjoined me to take at least 20 seconds to bring the speed up. 'Gives the chaps in the boiler room time to get the extra burners in,' he said. As the propeller started to bite and accelerate the massive ship, we could feel the hull starting to jump up and down a bit, rather like an old car being accelerated too hard and getting axle tramp until the wheels got a grip. This only lasted a minute or two, and then everything became quite smooth once again, as the ship's speed caught up with the propeller.

We maintained Full Ahead for about 20 minutes, before getting a Slow Ahead movement for another 10. 'That'll be the pilot getting off,' said JH. The next movement was Full Ahead again, and then the telephone rang. This was the bridge to say that they would ring Full Away at 2000 hrs – in about 5 minutes' time. Spot on 2000, the telegraph clanged right round from Full Ahead to Full Astern and back again, which was our signal to record the propeller shaft counter at the start of the sea passage and begin increasing speed.

The counter reading was always recorded at the end of every watch and the total number of turns of the shaft worked out and recorded in the log. This figure multiplied by the pitch of the propeller would give the Engine Distance Travelled, which was communicated to the bridge every day at noon. If we were too far away from land for the navigators to get any direct bearings and they hadn't been able to get any sun or star sights, either, to work out the ship's position, then the engine distance and the course, with allowances for the current and propeller slip, would give them a dead reckoning position. When I started to get interested in the navigation side of things I was surprised to find how small the slip figure was – usually less than 10 per cent, which shows how efficient the propeller and hull shapes were.

I should explain at this point that Full Ahead for *Yorkshire* under manoeuvring conditions was 75 rpm, whereas Full Away meant the actual maximum speed that had been decided upon for the voyage, which in this case was going to be 105 rpm. It would take nearly an hour to build up the final 30 revs, as various systems that had been using live steam from the boilers when under harbour and stand-by conditions had to be changed over to steam bled off from the turbines, this being more thermodynamically efficient. The burners in the boilers also had to be changed one by one, for another set with larger tips, to cope with the increased steam demand. The electrical officer (always called Lecky on board ship, as opposed to the radio officer who is Sparks), meanwhile adjusted the loads on the gennies until one of them was taking the lot, whereupon he tripped the circuit breaker on the other, which could then be shut down. As we had now finished with steam on deck for the winches, the auxiliary condenser, with its associated pumps and air ejectors etc, could also be shut down.

I stayed down below for the extra hour to see all this being done, so that when I finally got back on deck it was getting dark and the Welsh coastline showed but dimly behind us, with just a few lights still to be seen along the coast. I stood for a while right aft on the poop, observing the writhing and foaming wake streaming out behind us. This was one place where there was always a lot of vibration to be felt, and standing there under the jackstaff with the Red Duster flapping in the breeze above my head, I felt a mixture of emotions: there was some pride that I was going to be part of the team that produced the thundering power I could feel underfoot, and also a touch of homesickness. I had never even been abroad on holiday before and now it was going to be three or four months before I saw my parents again – not to mention my cat, Spud, who had shared my bed nearly every night for the previous ten years.

After a shower, I got changed and went back aft again for a beer in the bar before turning in. I was chatting to Fred, who was complaining that he had been with JH for the previous trip as well and had yet to be given a go on the throttles, which I thought was a bit rough; but he bore me no grudge and was to become one of my best mates on *Yorkshire*.

After the usual rude awakening next morning, I paid a quick visit to the bridge to see where we were before going down below. The second mate on watch pointed to an X on the chart, and a walk out onto the bridge wings for a thorough look round in all directions showed just one light, way over to port – Bishop Rock lighthouse just beyond the Isles of Scilly, I was informed. Coming off watch again four hours later, I could see nothing at all – just a slate-grey ocean and an empty horizon in all directions. I really was at sea now.

5 A VOYAGE TO THE PERSIAN GULF

Yorkshire was very generously manned even by the standards prevailing in the 1960s. On the deck side, after the captain there was a chief officer (who did not keep a bridge watch but was responsible instead for all cargo operations and the deckside maintenance), together with the first mate and two second mates, who *did* keep the watches. On the engineering side there was the chief engineer, senior second engineer and an extra third engineer who were all on day work, while the watches were kept by the junior second, third and fourth engineers, each with a fifth engineer to assist them. We also had a radio officer, an electrical officer and the chief steward or purser.

The petty officers comprised the bosun, who was the senior deck rating, a carpenter and a pump-man who were all under the charge of the chief officer, and in the engine room a donkey-man. Historically, the donkey-man was the man left in charge of the auxiliary (donkey) engine when a vessel was in port, although on *Yorkshire* he was simply the senior engine rating, and worked as required. The carpenter, incidentally, did precious little woodwork and was mostly engaged in general repairs on deck and around the accommodation, although he also took a set of soundings of every single tank or void space every day, which he presented to the chief officer for the noon log. The chief cook also ranked as a petty officer, while the ratings included about six seamen, three greasers, a second cook, two stewards and a galley boy. Finally, we had a deck cadet and myself, which made a total ship's complement of 37 men and the 2 wives I mentioned previously.

It did not take me long to discover that it was not the captain and chief engineer who ran the ship, but the chief officer and the second engineer – on *Yorkshire* it was not even the senior second either but JH, who was by far the more dynamic of the two characters and was always about after breakfast organising the plan of action for the day workers. The chief officer did something similar for his set of men, which for the first few days of the voyage, at least, was going to be washing and gas freeing the cargo tanks, which went on day and night until the job was completed.

An oil tanker when it has just finished discharging her cargo is in its most dangerous condition: the 'empty' tanks are not of course really empty, being completely coated with oil residues and also full of flammable gas which has to be removed; this was before the days when the tanks could be backfilled with inert gas as the oil was pumped out, which made the procedure a whole lot safer. Tank washing then was carried out with a kind of giant lawn-sprinkler device called a Butterworth machine, which was driven by sea water at high pressure supplied from the engine room. Cold water is usually okay for the lighter grades of crude oil, but the heavier types will require it to be heated. Several of these machines were lowered down into the tanks through a series of openings on the main deck and given some time to work before being dropped down a bit further and so on until the tank was completely washed out. Next, gas freeing was carried out; this was done by water-operated extractor fans, which were placed over the same openings and sucked out the fumes. Handheld gas analysers were available to check the atmosphere periodically until a safe level was reached.

After tank washing was completed there would be considerable quantities of oil and water residues remaining at the bottom of the tanks. These 'slops' were pumped into the 'slop tank', which is usually the aftermost centre tank of the ship and then several days were allowed for the mixture to separate out and the oil to float up to the top. The water was pumped out of the slop tank into the sea, until the oil was reached, when the pumping would be stopped. There was some instrumentation to let the pump operator know when this oil/water interface was reached, but it was not terribly reliable and quite often it would be a case of simply observing the overboard discharge until it started to go black and then quickly stopping the pump. Regrettably, therefore, a barrel or two of oil usually went overboard at these times, although the quantity was so small that it was dissipated almost immediately by the wash from the propeller. When the next loading port was reached the fresh cargo would be pumped in on top of the leftovers.

Before this Load on Top system was introduced (back in the 1950s I believe), most tankers simply pumped all their tank washings overboard with consequent heavy pollution of the world's oceans. Even when this became an offence under maritime law, some flag of convenience operators still carried on doing it for a few more years when they were out of sight of land and thought they could get away with it. Finally, however, this menace was brought to an end by the activities of environmental campaigners, who named and shamed various companies by showing aerial film of their tankers leaving behind their tell-tale black trail. Apart from anything else, Load on Top actually saves quite a bit of cargo and therefore money, so it seems strange that it took so long for it to become standard practice.

All this furious activity on deck was of course nothing to do with me beyond being asked to start and stop the Butterworth pump and heater as required, because the first two or three hours of my watch in the mornings were usually spent crawling round the bilges tracing out pipe systems and valves as directed by JH. What made it more difficult was that none of the pipes were colour-coded as we had been told at college they might be. Every pipe below the floor plates was simply painted in red oxide and everything above was either painted white or covered in white lagging. The valve handwheels were coloured, but this was not an infallible guide to what was actually in a particular pipe – the seawater, bilge, ballast and Butterworth valves were all green, for instance, while quite a few of the smaller ones were not fitted with the little engraved brass plates that said what they were. At this stage of my training, even some of the ones that did have plates might as well have been written in ancient Greek for all the good they did me – AUX.AEJ.RET.TO ADT was one fine example: auxiliary air ejector returns to atmospheric drain tank.

The first system JH wanted me to record was the boiler feed water system, which was one of the most complicated but when fully understood would give me a pretty good grasp of how the plant actually operated. I was to start at the main condenser and work my way along through all the various heat exchangers and de-aerator etc until I got to where it was fed back into the boilers.

My first attempt was a simple block diagram showing what I thought were all the basic components and the order in which they came up. This took me a couple of watches before I submitted it for his approval. 'That's complete bollocks is that – go and have another try, and this time put in some valves,' was his first comment. I went back for another round of bilge diving and scrambling round the boiler room, and showed him the Mark 2 version on the following day. 'That's still crap – don't you know that every heat exchanger has an inlet and an outlet for the stuff being heated or cooled, and the same for the stuff that's doing it?' Back I went again, this time filling in all these other pipes as well, which took another watch. He spent a bit longer looking at this third version, but it was still rejected. 'That's better – but you've missed out the make-up feed and spill arrangements (these are supposed to maintain the total amount of feed water in the system at a constant level).' Having rectified these omissions, I had high hopes for my latest attempt but it was still not good enough for JH. 'Right; now all you have to do is put in the temperatures and pressures at each stage of the system and then you're done!' I put these in as requested, it got past his eagle-eyed scrutiny, and I was finally allowed to make a fair copy of it in the notebook.

After the feed system, I had to record the steam system, seawater cooling system, bilge and ballast systems, fuel and lubricating oil systems, compressed air system and several more besides. Although I had by now a pretty good idea of what sort of standard he expected, each of my drawings usually took two or three tries before it was accepted. After a few weeks of this I hated him with a passion – but it was tempered with respect, because although he could be a real bastard and treated all his subordinates like dirt, there is no doubt that he was one of the cleverest engineers I ever met, and I probably learnt a lot more in a shorter time from him than I would have done with anyone else.

For the final hour of the watch, I was expected to get in the engine room log. When Fred did it he could have it completed in about 30 minutes, but when I first went round, I spent an hour – and still hadn't finished when the 8 to 12 came down to relieve us. I gradually got quicker as I worked out the most logical route and the best angle to view the various thermometers and gauges from – some were placed so you had to pretty well hang upside down with your head below the floor plates to see them. There was a lot to do, and of course I made the occasional mistake, which, as the log was completed in ink, stayed there for all to see. JH took a dim view of my crossings-out and gave me a ticking off in his usual aggressive manner, which of course made me even more nervous. Possibly because of this, on the very next watch I made three mistakes on a single page, which led to a most fearsome dressing down in front of the other watchkeepers – not helped by the fact that JH had a bad case of halitosis, which was particularly unpleasant when he was yelling in my face from only a few inches away.

After this, of course, I was so petrified about making a mistake I could hardly think straight, with the result that the following morning I somehow managed to get a whole column of figures for the port boiler transposed with those from the starboard. When I saw what I had done I was aghast. I thought perhaps to just leave it as it was and hope he wouldn't spot it but finally decided that as this was extremely unlikely, it might be better if I made the corrections first rather than have JH find them, so I crossed out the whole lot as carefully as possible and then wrote them all in again in the right place. When JH saw the book he went ballistic, saying, 'You did that on purpose, you little bastard! I'm going to have you up in front of the chief for this.' He then told me that after the usual morning field day I was to put on my No.1 uniform and report to the chief's cabin just after noon, which was when the log was taken up to him to be checked and countersigned.

I duly got fully booted and spurred and arrived outside the chief's cabin at the appointed time, where I met JH, also in full uniform and with the offending logbook under his arm. I have to admit I was quaking in my boots as JH knocked and went into the chief's office; I had no idea what punishment might be in store for me, although I guessed that being clapped in irons or made to walk the plank were probably no longer allowed.

The chief was sitting at his desk and JH put the log book down in front of him, and proceeded to point out all the mistakes and tell him about my general bad attitude. The chief took some time looking at the book without saying anything, and it gradually dawned on me that he was drunk. His head moved slowly in and out as he tried to focus, and when he eventually spoke it was simply to ask JH what the time was and upon being told, whether he would like a drink. When JH declined this offer with a very exasperated expression, the chief seemed to notice me for the first time and asked me who I was. When I told him he asked me if I wanted a drink. I couldn't see that it would make things any worse by accepting, so I said, 'Thanks very much, Chief, I wouldn't mind a beer.' By this time JH was pretty well incandescent and he stormed out saying that he had better things to do and that I was to see him again after lunch, which sounded rather ominous. In the end, however, my punishment was simply to have to work all afternoon as well as my field day in the morning, which on balance I felt was quite lenient. Strangely enough, JH never mentioned the incident again, although he still rode me very hard and seemed to delight in finding me jobs way down in the bilges, or else in the heat at the top of the boiler room.

The above events took place over several weeks, during which time the ship had been making her way steadily south at around 16 knots. By the second morning it was warm enough for me to start wearing my tropical shorts and shirt, and I never put on the No.1 uniform again until we were back in the UK. A few days later I came off watch just after 0800 one morning and instead of the usual empty horizon, I was astounded to see a perfect, conical, snow-capped mountain rising from the sea off the starboard bow. This was El Teide, the massive 12,000-foot dormant volcano on the island of Tenerife. We were still so far away from this famous maritime landmark that its base, which was surrounded by low cloud, could not be distinguished from the horizon, making it look as if the mountain was suspended in mid-air.

A few hours later we were tied up in Las Palmas on the island of Gran Canaria, and Fred and I stepped ashore for a brief sightseeing tour – my first ever foray onto foreign soil. We hadn't time to do much, so settled for a stroll down the quay into town, where we found a bar and I was able to buy Fred a beer for a change.

The following morning we were off again, having taken on some 9,000 tons of heavy fuel oil, which would be enough to see us to the Gulf and back again. Down below I was doing the log as usual, so Fred finally got to have a go on the throttles, which seemed to please him immensely. 'Can't call yourself a marine engineer until you've worked the engines,' he commented.

For the next ten days we continued south towards Cape Town, where we were going to stop for mail and fresh supplies, and I learnt the meaning of the old song 'I joined the navy to see the world, And what did I see? I saw the sea!' Although we must have passed quite close to Cape Verde, the most westerly point on the African coast, I never saw a thing. Every time I went up and down the flying bridge or stood on the poop deck watching the wake, there was always just an empty horizon to be seen, and I marvelled at the determination of the early explorers in sailing ship days, who kept heading out across the open ocean on a hunch or hearsay, often for months at a time, in the hope of sighting some uncharted and unknown land.

During this leg of the voyage we crossed the equator, and the chief steward organised a Crossing the Line ceremony to mark the occasion. After lunch, those crew members who had not been across before were summoned to the swimming pool for their initiation. The captain was dressed up as King Neptune, attended by several of the other officers also in fancy dress acting as his courtiers. First off, he read out some mumbo-jumbo about the dangers and privileges of being admitted to his realm, after which the novices were called up one by one to stand in front of him to swear an oath of allegiance that acknowledged King Neptune as their lord and master. The chief steward had made up a bucket of some disgusting-looking multicoloured slop from the galley and after the swearing in, all of us, including the two wives, were liberally daubed with this muck from head to foot before being thrown into the swimming pool. After this we were issued with certificates to prove we were official citizens of Neptune's Kingdom, which were supposed to spare us from any further such ceremonies – they never did, though! Following the official business the occasion developed into a bit of a party as the chief steward had provided plates of nibbles, while there were free drinks as well – all paid for from the profits of the bar. With the sun blazing down and the pool available to cool off in when required, it was a very pleasant way to spend the afternoon, and the jollities carried on right up until dinner time, so that the 12 to 4 had a chance to join in as well.

The swimming pool was seawater filled and was not particularly big – about 15 feet by 10, and 6 feet deep I seem to recall – although the depth very much depended on how much the ship was rolling; in heavy weather it could get as low as 3 feet, because of the amount that slopped over at the ends when the ship rolled. It was situated behind the aft accommodation block, one deck up from the poop, and was a favourite spot for sunbathing and enjoying a beer in good weather, which seemed to be for most of our trip south.

Now the pump room was gas-freed, Fred and I were given the task of repairing the stripping pump during our usual morning field day. First of all I met Fred in the workshop and started to collect a bucketful of spanners and whatever other tools I thought might be required. Fred stood back watching me for a while, looking quietly amused, before asking me what I needed that lot for. I replied that I thought we were going to strip down the stripping pump, which would require some spanners would it not? Fred thereupon declared, 'We don't need all those,' and simply picked out a 2lb hammer and a hefty cold chisel. I was rather puzzled as to the nature of the repair Fred thought he would be able to carry out with these, but if it saved me the job of lugging my heavy tool bucket all the way up on deck, then down to the bottom of the pump room and back again, it was okay by me. I duly followed him down into the gloomy and claustrophobic depths of the pump room, which may have been gas-freed but still smelt pretty strongly of oil to me.

Arriving at the pump, he said that it was nearly always the shuttle valve sticking that stalled these pumps and he showed me how easily he could undo all the nuts on the valve chests of the pump with his hammer and chisel. I didn't like to mention that it mangled them as well but the bottom line was that we had the shuttle valve stripped out, cleaned, freed off and the whole lot boxed up again inside an hour – the hammer and chisel proving to be equally efficient at doing the nuts up again. After this we tested the pump and had the rest of the morning off. Once again, I had learnt some new tricks.

The fine weather continued right down to Cape Town, which is one of those places where what you actually see is better than any picture postcard view. *Yorkshire* was far too big to go into the harbour, but even from where we stopped a mile or two offshore, the view of Table Bay with Table Mountain behind was simply fabulous. Unfortunately our stay was limited to about half an hour, which was all it took for the launch to come out to us and deliver our fresh supplies and the all-important mail, following which it was Full Away again. Our next stop would be Kharg Island in the Gulf.

For the next couple of days we remained in sight of land as we continued around the bottom of Africa, passing Port Elizabeth and East London. The main claim to fame of these waters is that back in 1938 someone caught a live coelacanth here – a species of lungfish, growing up to 2 metres long, previously believed to have become extinct at the same time as the dinosaurs. Then we bore away to the north-east and into the Mozambique Channel. This stretch of water, which lies between the mainland of Africa and the island of Madagascar, has an evil reputation amongst seafarers, with many tales of freak waves and vessels disappearing without trace. As we passed through it, however, the sea was quite placid and the most incredible translucent blue colour I had ever seen – almost like a child's painting where they have used the cobalt blue straight from the paint-box without toning it down.

The next land we saw was the Comoros Islands at the northern end of the channel, which we passed by close enough to be able to see a few settlements with simple huts thatched with palm fronds and smoke rising from their cooking fires – a scene that possibly had not changed since the first European explorers had come this way some 500 years before. Now, I am sorry to say, an airport has been built on the main island, and the Comoros have become yet another unmemorable package holiday destination.

As we progressed further north into the Indian Ocean, it grew steadily hotter, with the temperature at the control platform now exceeding 100°F. At the top of the boiler room, where the washing was hanging up, it was over 130. What made it bearable was that there was plenty of ventilation in most areas and water coolers were provided in both the engine room and boiler room, so you could always get a drink of cool water. The accommodation was air-conditioned, however, and when I got back to my cabin, where the temperature was kept at about 75°F, it felt positively chilly.

Although I had by now completed all my system drawings to JH's satisfaction, he still kept finding things to test me out on, and would ask me questions about something I should have noticed (but probably hadn't) during my walk round the job each watch. On one occasion the question was 'Which was the hottest bearing you went past today?' I thought hard for a minute and then guessed it might be on one of the forced-draught fans, simply because they were high up in the boiler room where the ambient temperature was pretty high anyway. So I piped up with 'The starboard forward FD fan bearing.' He gave me a quizzical look before replying, 'I guess I shall have to give you that.' One up to me!

One of his favourite tricks was to point at some insignificant valve without a label on it and ask me if I knew what it was, which of course I usually didn't. His next comment would always be, 'You'd better bloody well find out, then!' I can only remember one occasion where he actually told me himself what one of them was for, this being a small green valve on the forward engine room bulkhead. Having asked the question and received my usual shake of the head, he said, 'That, my ignorant little friend, is the sanitary water supply to the forward accommodation, without which our glorious captain will be unable to flush his bog!' This was one of the few occasions when we had a laugh together.

By now we had reached the Arabian Sea, the north-western part of the Indian Ocean, which extends into the Gulf of Oman. The passage from there into the Persian Gulf, the Strait of Hormuz, is a narrow and tortuous channel, and right in the narrowest part there is a group of small rocky islets called the Quoins (pronounced coins). As we passed through here into the Gulf, it was getting late in the day and the sun was catching these little lumps of rock at a low angle, so that they appeared to be glowing red hot. Behind them, on Iran and Oman on either side, range after range of hills fell away into the distance, and as there were no buildings or vegetation of any kind to give a sense of scale it was impossible to tell how high they were. The scene was beautiful and also quite surreal – the sort of landscape you might expect to see on some strange planet in an episode of *Star Trek*. As the sun sank still further, purple shadows crept up the sides of the hills while their bases grew steadily darker until they merged into the sea, which had also changed colour from the deepest blue to an inky black. Finally, only the highest summits remained in view, suspended in the sky like a row of jagged teeth, stained blood-red.

After this dramatic entry into the Gulf, we had nearly another 18 hours of steaming before arriving at Kharg the following afternoon, and if I thought it had been hot in the Indian Ocean, it was nothing compared to the heat in the Gulf. This stretch of water is almost totally enclosed by some very hot landmasses and also is not particularly deep – around 30 fathoms on average – which means the whole of it can get very warm. In summer, the sea temperatures are often around 90°F; in fact I have recorded 93°F on more than one occasion. The engine room temperature was now 115°F on the plates, while at the top of the boiler room it was a staggering 145°F (63°C) – we could hang a soaking wet boiler suit straight from the washing machine over the handrails up there and in half an hour it would be bone dry and stiff as a board. Fortunately, the only reason we had to go there, apart from hanging out clothes, was to record the uptake temperatures for the log and to take flue gas samples for testing, which would only take a few minutes. One thing I found out pretty soon, was that any metal in contact with the skin at these temperatures would burn it, so gloves had to be worn to make it possible to grab the handrails, and my watch and my St Christopher on its chain had to come off.

Needless to say, under these conditions the water coolers got plenty of use, and at mealtimes there were also jars of salt tablets available, which we were advised to take to avoid getting heat exhaustion. With all the fuss made these days about the dangers of having too much salt, it is perhaps surprising that although we would all be pouring salt liberally over our food and taking possibly a dozen salt tablets every day as well, none of us ever seemed to suffer any ill effects. If you didn't take enough salt, however, a collapse from heat stroke would soon follow, which could be quite serious. When off watch, cold beer seemed to have the most marvellous restorative qualities, such that most of the engineers seldom managed to make a case last much more than a couple of days.

Kharg Island, I am afraid, is never likely to make it into the top ten list of favourite ports of call for any seafarer, consisting as it does of just a few square miles of bare brown rock with not so much as a blade of grass to add a touch of green. On one side, a pier extends out into deep water, and at the end there is a long jetty, capable of handling two ships the size of *Yorkshire*. The island is connected to the mainland by an underwater pipeline and it used to be the main oil export terminal from Iran until 1986, when Iraq bombed it almost out of existence, although I believe it has since been rebuilt and is operational once again.

Having a couple of hours to spare in the afternoon before our watch, Fred and I decided to take a stroll down the pier to the seamen's club, which we had been told was at the end of it. We were frisked by an armed Arab security man before going down the gangway to ensure we weren't carrying any matches or lighters, after which it was a ten-minute walk in the ferocious sun down to the club. This turned out to be a very sad little place with just a few tables and chairs and a bar selling soft drinks and non-alcoholic beer. There was also a billiard table with a big tear in the green baize, and a battered dartboard but no darts. Outside there was a courtyard with a few more seats and, quite bizarrely, a collection of stone animals – lions, elephants and camels amongst others – obviously it had been designed as a children's playground but it was situated in a place where children were never likely to be.

We were not allowed to walk any further than the club to explore the interior of the island, and indeed we felt little inclination to do so as it was about 120°F in the shade, so Fred and I settled for a couple of Cokes in the bar before leaving the concrete zoo behind us and walking back to the ship – but not before another unsmiling Arab with a gun searched us for anything inflammable. The water below the pier was very clear and we could see to the bottom until we were at least halfway to the jetty, at which point it must have been 30 feet deep. Shoals of small fish could be seen darting in and out of the steel supports, with much bigger ones, possibly sharks, making an occasional appearance.

Arriving back at the ship, which already seemed to be a few feet lower in the water as she filled up with finest Iranian crude, we went back to my cabin for a change and had a proper beer before it was time to go below again. At this point I can hear you 'tut-tutting' about the dangers of drinking before going on duty, but I can assure you that in the temperatures in which we were working, the effects of one beer would have been sweated out almost before we got down the ladders to the control platform, so I make no apologies.

The ship was fully laden in less than 24 hours, and we were on our way back home – and I am sure that none of us would have been particularly sad if we never saw Kharg Island again. Another two days' steaming saw us back through the Straits of Hormuz, the Gulf of Oman and into the Indian Ocean once again, where the sea temperature was down to 85°F – this made things quite a bit more comfortable in the engine room, although still ruddy hot.

The secret of being able to keep going in these conditions was to drink plenty of water – probably around 3 litres per watch – and to ensure you had enough salt; if you couldn't taste it in your own sweat, then you weren't getting enough. The other trick was to take every opportunity to stand under one of the ventilation duct outlets (we called them blowers) and let the air blow down the front of your boiler suit – a minute or two of this was most invigorating, even if the air coming down from outside was 90°F or more. I also took to finding a suitable piece of cloth from the rag bag and folding it into a sweat band to tie round my forehead, which made me look rather piratical but at least kept the sweat from stinging my eyes.

By now I was quite used to the routine of life on board ship and was very comfortable in my cabin amidships. Like most of the single men on board I had put up a few pin-ups on the bulkheads to make the place more homely, but instead of the usual magazine centrefolds, mine were from a much more original source. The Tennent's export beer cans at that time were adorned by pictures of girls in bathing suits - and there were 24 different ones. By the time I had drunk my way through three or four cases I had the complete set, which I had cut from the cans with tin snips, flattened out and stuck on my bulkheads.

Around midday if the weather was fine and clear, the deck officers could be seen gathering on the bridge with their sextants to take the noon sun sights. This was long before the days of satellite navigation, and everything was done in a way that would have been familiar to Captain Cook over 200 years before. I was fascinated by astronavigation, and would quite often go to the bridge myself to observe the process. A sun sight would be taken when the sun was at its zenith (highest point in the sky) and this would give the ship's latitude. It was not very accurate for longitude, however, because the angle the sun makes with the horizon remains pretty much the same for several minutes as it passes the zenith. So the exact time of the zenith is hard to judge – and being wrong by just one minute would mean an error of 15 nm at the equator (an hour, being 1/24th of a day, represents 15 degrees or 900 nm of the earth's circumference at the equator (21,600 nm), so one minute of time equals 15 nm). The time the sun crosses the Greenwich meridian also varies by several minutes either way over the course of a year, and this could introduce another, even larger, error, although corrections for this can be found in the Nautical Almanac.

So to obtain longitude as well, star sights, or occasionally planet sights, would be taken at twilight in the morning and evening; it was no good trying it after dark, even though the stars might be burning bright, because the horizon would usually be invisible or indistinct, and it was the angle between the star and the horizon that was needed. Some sextants, especially those used in aircraft, have an artificial horizon given by a bubble in a kind of spirit level device, but none of our deck officers had one of these. As it started to get dark, then, the mates would be on the lookout for the first stars to appear, and as soon as one had been positively identified the sight would be taken and a stopwatch started. The officer would then go to the chart room, write down the reading from the sextant, read the chronometer and subtract from it the time recorded on the stopwatch. This process would then be repeated until he had shot at least three stars. Then, by means of the astronomical data in the Nautical Almanac, the sight reduction tables and some maths, a position line could be obtained for each sight that would be marked on the chart; if accurate, the ship would be somewhere along this line. One position line is not enough to get the actual position of the ship but two will give a fix at their point of intersection. Three such lines obtained from stars in different parts of the sky are better still, and when plotted will form a triangle where they intersect – the ship will then be somewhere in the vicinity of this triangle. If it was a small triangle – say 2 or 3 miles across – then that was a pretty good result, and quite accurate enough when you were in the open sea.

Many of the stars that the navigators chose were ones that I had never heard of. Arcturus, Altair, Fomalhaut, Spica, Procyon and Capella were some I can remember from those days, the strange-sounding names often having an Arabic origin, as it was the Arabs who were the first proper astronomers. Ideally, three widely spaced stars would be chosen – it would be no use picking three close together, as their position lines would then intersect at very shallow angles, giving a long thin triangle from which it would be difficult to estimate an accurate position. As well as data for the sun and moon, the Nautical Almanac lists similar information for the brightest planets (Venus, Jupiter, Saturn and Mars) and no less than 56 stars, so it can be seen that one of the skills of the navigator is to be able to select the ones that will be useful at a particular time of day, and of course to be able to recognise them as they appear at dawn and dusk.

All this 'white man's magic' is now pretty much a thing of the past, thanks to the advent of GPS navigation systems (global positioning by satellite), where even a cheap handheld device can give a position that is accurate to around 50 metres – at any time of day or night and in any weather. This saddens me in some ways, as we seem to be becoming more and more dependent on technology and less on our own skill and resources. The British Merchant Navy still insists on some knowledge of celestial navigation to pass the master's certificate, but in everyday use the black box, which nobody apart from its designers fully understands, has triumphed.

Occasionally in the evenings, instead of going to the bar I would go up to the radio room and play chess with Sparks. He was not actually employed by Esso but worked for the Marconi company instead, which had also provided the radio equipment. He was required to maintain a listening watch on the distress frequency at certain times of day – and these times were always in GMT. There were two clocks in the radio shack, one showing ship's time, while the other kept GMT and had a number of red and green sectors marked on it, which showed the required listening hours on the two main distress frequencies.

While we played our game, the receiver would be tuned in and Morse code could be heard dah-dah-ditting away in the background at a whole range of different pitches. How he could play chess, hold a conversation with me and pick a message out of the air at the same time was quite beyond me. Occasionally something would catch his ear and he would grab his pencil and note it down; as often as not this would turn out be some rubbish that should not have been transmitted on the distress frequency anyway. We did receive genuine distress calls on rare occasions, but mostly they would be from locations too far away for us to give any assistance, so the call would simply be logged and no further action taken.

When we were approaching port we had VHF radio for voice communication with the harbour master and pilot etc, but the range was very short, so nearly all messages between ship and shore were passed in Morse code. As our ship sailed under a British flag, all our messages were routed via Portishead Radio, which is situated on the Bristol Channel. Twice a day, I think it was, Portishead would transmit a traffic list, which was just the four-letter call signs of any ships that had messages waiting for them. If we had any, then Sparks would wait his turn in the queue, getting ready to write them down, together with anything we might have wanted to send back. It was all a bit of a black art which depended on being able to bounce the radio signals off the ionosphere to get them round the world, and a sparks had to know or guess the frequencies that were likely to work in certain parts of the world at different times of day.

The equipment looked very dated even then, with the main transmitter being about the size of a small wardrobe, with large Bakelite handles and knobs to set the frequencies and a meter to prove that something was actually being sent up the aerial. However crude it might have looked, though, the system worked, and it would not be superseded for at least another 20 years on most ships.

As we reached the southern end of the Mozambique Channel, the weather worsened and we started to take a few seas over the decks for the first time on the voyage

– much to the disgust of the chief officer, who was trying to get the painting of them finished. The freeboard was of course very much less than it had been on the outward trip and with our total weight now well in excess of 100,000 tons, the ship did not exactly ride the waves but ploughed straight through them instead. When a bigger one hit us, you could feel quite a shudder as we slammed into it, and a second or two later the bridge windows would be obliterated by the spray flying back from the bows. With the ship making 15 knots and the waves probably doing roughly the same in the opposite direction, when one of them was high enough to get over the side it came on board like an express train, sweeping down the deck until it hit an obstruction like one of the tank hatches, when a column of spray would be shot high into the air. Regrettably, I never managed to get any decent photographs in heavy weather because when it was really rough there was just so much spray in the air that I could never keep the camera lens dry for long enough to compose a shot.

I have to say that I enjoyed my first taste of rough seas and thought it was great fun trying to dodge the waves and avoid a soaking as I made my regular trips up and down the flying bridge. *Yorkshire* was of course a very large and well-found ship, and would shrug off a force 8 or 9 gale – it was to be quite some time before I was to learn (by being scared half to death, actually) that the sea could be dangerous, even to vessels as large as this.

Apart from the bar, the main social activity on board was a twice-weekly film show. We had a 16mm film projector and sound equipment (no videos or DVDs back then) and it was usually the cadets' job to set it up and show the films. The usual time was 2030 hrs, after the 4 to 8 came off watch, while the 8 to 12 had their own show after lunch. The films themselves were provided by an organisation called Walports, which kept stocks at most ports throughout the world. The films came in aluminium boxes, usually with three films per box, and whenever the ship docked the boxes would be exchanged. If it was the agent that was providing them, as at Cape Town where we didn't actually tie up, you had to take whatever you were given – and sometimes it could be real rubbish. On the other hand, in some places it was possible to visit the store and pick out a decent selection. Most of the films had two or three reels, which gave everyone a chance to get a fresh beer from the fridge as they were being changed on the projector. Some epics such as *Cleopatra* or *Ben-Hur* would even have 4 or 5 reels, in which case there might only be two films in a box.

If there was ever anything in the way of bare female flesh to be seen, there would usually be cries from the audience to rewind that bit and run it through again (the wives on board soon learned to accept this as normal). I can remember one particularly juicy scene from *Lolita* or something similar, where that section of the poor old film had been through the projector so many times, that the perforations in the edge where it engaged with the drive sprockets were quite worn out, which led to much booing and hissing, as it kept jumping the teeth and refusing to project properly.

Cape Town the second time round was no less impressive than the first, after which we were off on the final leg of our homeward trip. We also now knew where we were going: it was to be Fawley near Southampton, Esso's main refinery in the UK. Uncertainty over one's destination was quite common on tankers, where you would be sent wherever the oil could be bought the cheapest, so that one's departure instructions from Europe might simply be 'Cape Town for orders'. The port of discharge could similarly remain unclear, often until we were as far north as the Bay of Biscay, owing to the difficulty of predicting which refinery needed a delivery the most and what the berthing availability was. The only thing we could be sure of was that the company had no intention of ever leaving us anchored up for a day or two while they made their minds up, in case we might be tempted to take a lifeboat ashore and actually enjoy ourselves.

Our passage north was quite uneventful, and a fortnight or so later found us rounding the Isle of Wight and entering Southampton Water. We tied up at Fawley in the afternoon when I was on watch, and about half an hour later there was a call from the bridge to say that Mum, Dad and brother Peter were on board, which came as a marvellous and unexpected surprise. JH, who had taken the call, also came up trumps and said I could have the rest of the watch off ('You'd better bugger off topsides!' being the actual phrase he employed), which was decent of him. I needed no second bidding and quickly dashed up forward and met them in the bar where Sparks had already fixed them up with drinks, which was also pretty decent I thought. We repaired to my cabin and had a wonderful few hours swapping news and stories until it was time for them to drive back home.

When *Yorkshire* left Fawley, our next voyage was to the Tripoli in Lebanon and back again, but as I have already devoted three chapters to this ship and didn't go ashore there anyway, I think it is time to move on. After all I have said about JH previously, you might think that I would have been glad to see the back of him, but in the end this was not the case: he just had very exacting standards and if you didn't measure up then you got it in the neck. I learnt a tremendous lot about the operation of steam plant from him, which left me much better prepared for my subsequent ships than I might have been if I had been with someone who was not so tough. I am sure that beneath his abrasive manner there was a decent bloke trying to get out, if he could have only allowed himself to lighten up a bit.

6 STEAM LEAKS AND CONDENSER TROUBLE

Esso cadets were expected to work five-month trips and as I had only done three on *Yorkshire* I was not expecting much leave. Sure enough, a letter arrived a week or so later instructing me to report back to Fawley to join *Esso Warwickshire* and serve out the remaining two. *Warwickshire* was going to refit in Amsterdam and the cadet training officer had thought that would be good experience for me. I was quite pleased as well, because I had expected there might be a few chances to go ashore before we sailed for the Gulf again.

Warwickshire turned out to be very similar to *Yorkshire*, but just a bit smaller and had been built in Germany. This time I had a cabin down aft with the rest of the engineers, and would be spared all the trips up and down the flying bridge. My boss on this ship was the senior second engineer, whose name, I think, was also John; he seemed like a decent chap, quite young once again and with a big black beard. He asked me a few pointed questions to find out what I had learnt on *Yorkshire*, one of which I remember was, 'Give me five reasons why you could be losing vacuum on the main condenser.' Thanks to JH's strict tuition I was actually able to think of eight (see Appendix 1), which impressed him considerably, whereupon he told me that I would be put on the 12 to 4 watch with GW, the third engineer to 'help him out a bit'. I was to learn that this really meant providing some back-up for the junior on that watch because GW himself usually took a couple of hours every night to recover from the previous evening's drinking session.

GW was what was commonly known in the Merchant Navy as a professional third – that is, an engineer who never obtained a second-class certificate and could not therefore progress beyond third engineer. There were a number of reasons why this might have happened: quite often they had entered the profession via the trade apprenticeship route and might well have left school without any qualifications. In these cases, going back to college in their twenties to start learning all the necessary theoretical subjects for their second's ticket might prove very difficult. Other individuals quite simply did not like the idea of the extra responsibility they would have to shoulder if they reached the higher ranks, and were happy to simply remain as third engineers.

GW fell into the first group, I think, and was a rather tragic figure. I seem to remember that his wife had left him, after which he had slowly succumbed to the demon drink. He was only about 5 feet 6 inches tall, weighed around 8 stone and seldom appeared to eat anything more than soup and a bread roll for dinner. He had also lost two of his front teeth in some shoreside brawl, which left him with a very gappy expression when not wearing his false teeth. For all that he was a good shipmate and was always around whenever there was a major work-up in the engine room – and we were to have plenty of those on the coming voyage.

The greaser on our watch was another interesting character: he was a grey-haired Yorkshireman of around 60, and was usually rather taciturn. I found, however, that if I ever asked him something he could be most forthcoming and helpful. The first time I came down on watch, he asked me how I liked my tea and then spotted the Saint Christopher hanging round my neck. 'What's tha wearing that for, lad?' he asked. 'My mum reckons it might save my life if the ship sinks,' I replied. To this he commented, 'Let me tell thee summat, lad: I was torpedoed three times during t'war and I saw scores o' them buggers go down!'

Our trip to Amsterdam was uneventful, and as soon as we were tied up in the NDSM (Nederlandsche Dok en Scheepsbouw Maatschappij) shipyard, an army of yard fitters came aboard and started dismantling various bits of our machinery. The main reason for our visit was to investigate the thrust bearing on the high-pressure turbine, which had been running hot. At the same time, the main reduction gearbox and various other items of auxiliary equipment, including some of the main superheated steam pipes were to undergo their five-yearly Lloyd's inspection (every major component in the engine room and all the shipside valves had to be opened up every five years under what was known as the Lloyd's continuous survey programme, in order for the vessel to remain insured and classified 100A1 at Lloyd's).

When all the lagging had been removed from the turbine, all the bolts holding down the cover had to be removed, after which the cover was carefully jacked up on four long steel studs, it being imperative that it was lifted up completely parallel to avoid damage to the blading. When it was clear, its weight was taken by one of the dockside cranes, whose hook had been lowered down through the engine room skylight hatch which had previously been removed for the purpose. The crane driver could not see anything that was going on down below, and all signals had to be relayed to him by a man stationed just outside the opening, and he in turn relayed those of the foreman down by the turbine – a mobile phone would have been very handy.

The turbine rotor was now removed with equal care, using the ship's own crane initially, after which it was transferred to the shore crane as before. The only problem was that on this occasion the shore crane hook was not directly beneath the end of the jib, which was at least 100 feet above it, so that when the load was transferred to it the rotor swung gently across the engine room like a giant pendulum until it hit a large steel girder. After the rebound it hung there for many seconds, ringing like an enormous tuning fork. Fortunately, high-pressure blading is very solid and has a thick steel ring riveted around the ends of each row of blades, so no serious damage appeared to have been done and the chief engineer, who had been watching the proceedings, could start breathing again. For me, it was the one and only time during all my years on steamships that I ever saw a main turbine with the lid off, which says quite a lot for the reliability of this type of plant.

With the boilers shut down and the ship running on shore power, we were not keeping watches while we were alongside, which meant that most of us were on day work and were therefore free to go ashore every evening – a pleasant change for tanker men. The only thing against this was that it was midwinter, and bitterly cold both on and off the ship, despite the large number of portable electric heaters which had been brought on board. For the most part I went ashore with Huw, the other engineer cadet, and we happily explored the sights and sounds of Amsterdam, finding, as well as the inevitable bars and cafes, a jazz club and a cinema showing English films.

After one foray, we got a bit lost on our way back to the ship and strayed into an unfamiliar part of town. We were surprised when we started to see girls sitting in what appeared to be small ground floor or basement shop windows. The first one we came across was very smartly dressed and kept absolutely still as we walked by – I thought at first she was a dummy until she winked at me. Such was our innocence that it was only when we got back to the ship and told some of the other lads about it that we were told that they were prostitutes.

By day six our little holiday came to an end as the boilers were fired up, and it was back to watchkeeping again. Esso was determined that the ship would sail after no more than a week, so toward the end everything was being done in a terrible rush to meet the deadline for departure. We were to pay the price for this haste before we even reached the open sea, as one main steam pipe joint after another failed simply because they hadn't been tightened up properly. The fitters from the NDSM yard who were responsible for this debacle were by now tucked up cosily in bed while we were left to redo their job for them – what we called them was quite unprintable.

If you imagined that a steam leak at 860 psi (60 bar) makes some kind of hissing noise you would be in for a very unpleasant surprise if you actually heard one, for I can assure you that it is not a bit like that. First off, you become aware of a clear and penetrating high-pitched whistle; then as the leak worsens, lower tones come in as well and it develops a singing quality of ever-increasing volume, rather like an enormously powerful choir where everyone is ignoring the conductor and doing their own thing. When it has reached this stage, the entire engine or boiler room, as the case may be, seems to be completely full of sound, so that it becomes impossible to tell where the leak is coming from. Finally, it develops into the most awful and terrifying screaming noise that is loud enough to be physically painful if you don't wear some form of ear protection.

Another unpleasant feature of superheated steam is that it is quite invisible, and great care needs to be taken when searching for leaks to avoid walking straight into one, with very painful results. The usual procedure was to tie a rag around the end of a broom handle and apply it to any suspect areas to see if the rag started to blow about. We had one other clue to help us find these leaks, and that was where the fresh lagging that had been wrapped round the joints started to blow off in places, so that for a time the air was filled with what looked like snowflakes falling. If I ever succumb to an asbestos-related disease, then I can be fairly sure that this was the day when I received the lethal dose.

As the ship was obviously in no fit state to put to sea we dropped anchor and commenced repairs – John put us all to working six hours on and six hours off until they were completed. Despite the fact that this was a very chilly November night, the steam leaks warmed the atmosphere and increased the humidity in the boiler room to an almost unbearable degree, so we had a most unpleasant job ahead of us.

The flanged joints on these superheated steam lines usually had 8 or 12 bolts something over an inch in diameter, while the joints themselves were of the Metaflex pattern, which are quite unlike anything you might find on a car. They consist of a steel backing ring like a big washer some 2–3 mm thick, which is of a size to just fit inside the ring of bolts. Attached to the inside edge of the backing ring is the actual sealing part, which is a spiral of very thin V-section stainless steel with compressed asbestos fibre in between the coils. The spiral seal starts out at around 5 mm thick, but once it has been pulled up hard by the bolts, the Vee sections will be flattened out, thereby compressing the asbestos to form the seal. They can only be used once.

As we had leaks on the superheater outlet pipework from both main boilers, we would have to shut down one boiler at a time and repair it as quickly as possible whilst putting up with the screaming racket from the leaks on the other one. The senior second gave GW and me this particular job, and split the rest of his men up into similar teams of two to deal with the others. We were lucky that both the superheater outlet valve and the next isolating valve down the line were both steam-tight, so that as soon as we had shut them our particular leak died away to nothing quite quickly: but before starting we still checked with the rag-and-broom-handle trick.

On the face of it, changing one of these Metaflex joints would appear to be quite an easy task – just remove half the bolts around a particular flange and loosen the rest, after which it should be quite simple to remove the old joint and slip in a new one. In reality it was not quite as straightforward as that: in the first place, the pipe had been at 500°C until we shut the valves on it, and as we hadn't got the time to hang around waiting for it to cool down we had to stand there on top of the boiler drum, getting roasted by the radiated heat and having to do everything whilst wearing welders' gloves. As we removed the bolts they were dropped into a bucket of cold water, where they hissed and sizzled furiously (they were quite big bolts and weighed over a kilo each, as I recall) – so by the time we got round to our second joint, the water was boiling and we had to get a fresh lot.

The other problem was that in quite a few cases simply removing the bolts would not result in the flanges springing apart far enough to get the joints changed, and we would have to rig a chain block to take the weight of the pipe so as to give us the required clearance. As soon as we had the required gap, I was all for swapping the joints as quickly as possible and boxing the job up again. GW would have none of this, however, and insisted we make a minute examination of the mating faces to ensure there was no evidence that the leaking steam had damaged them – if superheated steam is left to leak for long enough it will cut grooves in a steel face. 'It's bad enough standing here doing this once – we don't want to be repeating it later when the ship is further south and it's really hot,' was his comment.

Fortunately our faces had no steam cuts, so we could get cracking and reassemble them. To line up the flanges, much use was made of a podger initially, this being a steel bar tapered down to a point at one end, which we poked through a pair of bolt holes to lever the flanges into the required position so we could get the bolts in. After this we could tighten them up, taking great care to ensure we kept an even gap all round each flange. First we used a pair of ring spanners until we could go no tighter, after which one of the ring spanners was swapped for a flogging spanner (one with a ring to fit the nut on one end and a solid lump of steel forged onto the other, designed to be hit with a hammer). We then went right round each flange twice, one of us holding the ring spanner on the bolt heads to stop them turning, while the other held the flogging spanner on the nuts with his boot and whacked them up with a sledgehammer. We had no torque wrenches and simply kept going until they started to ring, which meant they were tight enough. Despite the crudity of the method, I can't remember us ever stripping a nut or bolt thread when doing one of these jobs, which says something good either about our judgement or the quality of the bolts.

After we had finished the first two joints in the boiler room we gently opened the valves again and let the steam through from the opposite boiler; much to our relief, they remained steam-tight, so we could flash up our boiler and isolate the other one, after which peace and quiet descended on the boiler room once again. John came to see how we were doing and was pleased that we had succeeded with our first two joints. He could also see that we were both drenched with sweat and generally knackered, so he sent us up topsides for a breather, which came as a very welcome relief. For me, it was a case of putting on a fresh boiler suit and throwing down a couple of coldies, but I noticed that GW chased his beers down with an enormous gin and tonic. As he never seemed to eat anything much I can only assume this was how he got the required calories to keep going!

After our little break, another long session on top of the boilers completed our share of the job and we could stagger off for a shower and a few blissful hours of sleep. When we came on watch again at noon the rest of the leaks had also been repaired and we were soon on our way again, much to everyone's relief.

One thing that *Warwickshire* had, which none of my other Esso ships did, was an engine control room; this was an air-conditioned box housing the main control console and the manoeuvring valves. It made a very pleasant change to have a cool place to retreat into when it was roasting outside. John was always quick to remind us, however, that 'The job's out there, not here in this box' and insisted that we spend a good proportion of our time outside, walking round the job, in the same way we would have done on any other ship. One other advantage of the control room was that we could hear any telephone calls a lot more clearly than we would have done in an open engine room, despite the fact that the phones there were always placed in supposedly sound-proofed booths. However, these booths were only just big enough to get your head and shoulders in and were open below waist height anyway, so when the machinery was running full out it was exceedingly difficult to hear any instructions or requests from the bridge. They must have got very fed up with me continuously asking them to repeat things.

On the 12 to 4 watch with GW I had my first lessons in water treatment for the boilers and the testing that was required to work out the dosages. The boilers on *Warwickshire* were of Babcock & Wilcox pattern and worked at 860 psi. At this pressure, water boils at 525°F (275°C) and the boiler water has to be kept pure: so pure in fact, that normal tap water would be no use at all – distilled water, with tiny quantities of chemicals added to prevent scale and corrosion, is the only sort that is good enough.

The problems arise from the fact that when the steam from the turbines gets through to the condenser it is cooled by sea water passing through thousands of small-bore tubes, which turns it back into water again – and it only needs one of those many small tubes to start leaking for the condensate to become contaminated with salt water. Once this contamination has got back to the boilers, it will stay there and steadily build up. This is because as the water boils off any impurities will be left behind – exactly the same as in a domestic kettle but on a grand scale. With the boilers each evaporating some 60 tons of water every hour, even a tiny percentage of sea water in the condensate would soon cause an unacceptable level of impurities.

Everyone knows that sea water contains plenty of salt (sodium chloride) but it also contains quite a few other dissolved compounds as well, in smaller quantities. The ones that most concern the marine engineer are those that cause scale to build up in the boilers, such as the sulphates and carbonates of calcium and magnesium. Sodium chloride is extremely soluble in water at all temperatures, but the scale-forming salts come out of solution at higher temperatures and will deposit themselves on the boiler's heating surfaces. Even a very thin coating of scale on a boiler tube will drastically reduce its ability to transfer heat to the water, so the efficiency of the boiler will be reduced and eventually the tube will overheat and rupture – very bad news in a boiler working at 860 psi. This is why we carried out a set range of tests every day to check the boiler water was okay.

Because sea water contains more sodium chloride than anything else (35,000 ppm on average), that was what we would test for when doing the boiler water checks. We would take samples from both boilers and allow them to cool. Then we would measure out the required amount for the test and add a few drops of potassium chromate indicator which would turn it a nice canary yellow. The actual test consisted of titrating in silver nitrate solution from a burette until the sample just started to change to orange, at which point we would stop and read the burette. Each tenth of a millilitre of silver nitrate we had titrated in equated to 7 ppm of chlorides, and for high-pressure boilers such as those on *Warwickshire*, we would hope to be getting readings of 35 ppm or less. If it reached 42 we would have to take steps to reduce it by opening a valve on the boiler (the blow-down valve) which would allow some of the boiler water to escape overboard. This would cause fresh distilled water from the storage tanks to enter the system and replace the lost water via the make-up feed valve. After GW judged we had let enough go, we would shut the blow-down valve and then test again, hopefully finding that the chlorides had now dropped to 21, which was considered to be quite good, or 28, which was okay.

We did a few other tests as well, including alkalinity, which was done by adding phenolphthalein indicator. This would stain the sample mauve if there was any alkalinity present, after which we would titrate in sulphuric acid until the solution went clear again. In fact we did this test first, because the same sample could then be used for the chloride test. The water had to have a certain level of alkalinity (30–50 ppm, I think), otherwise there was a risk of acidic corrosion of the steel boiler material.

Another important test was for phosphates, which we added to the boiler water to prevent scale formation. The check for this was to take a sample and add a reagent – can't remember what exactly this was – which hopefully would turn it a medium-blue colour. We would then compare our sample against a row of test tubes showing various different shades of blue with the equivalent phosphate levels marked under them until we found one that matched the sample. If this test, or any of the others described, produced results that were outside the set parameters, then we had to take a series of corrective actions as described a bit further on.

Apart from testing the actual boiler water, we would also check the feed water before it entered the boilers in case it had become contaminated with sea water from a leaky condenser tube. As the chloride level in the feed was normally extremely low (below 1 ppm), the normal test would not be sensitive enough unless we diluted the silver nitrate solution or worked with a bigger sample, so we normally depended on the salinometer instrument, which was mounted on the console. This worked by very accurately measuring the electrical conductivity. I can't remember the figures on the dial, but I know it was calibrated in three segments: green at the left, okay; in the middle was yellow – trouble brewing – and on the right it was red, which meant we had to do something about it fast before the boilers became badly contaminated. This dial was to get an awful lot of scrutiny during the return half of our voyage.

After we had done all the testing, we then had to work out what actions were required to keep the boiler water in the required condition. This could simply be adding further chemicals – sodium hydroxide to bring up the alkalinity, and some phosphate compound if required. These chemicals were mixed with water and poured into a dosing pot, which was connected by valves into the condensate line downstream of the condensate pump. The valves had to be shut, then a drain opened to empty the pot, after which the lid could be removed, the drain shut and the chemicals added. Then the lid was secured again and the condensate inlet and outlet valves opened, which sent the lot off on its merry way toward the boilers. One other chemical was added continuously to the boiler feed water by a metering pump – this was hydrazine which was supposed to remove any dissolved oxygen (steel cannot rust if there is no oxygen present). We had another colour comparator test for hydrazine (a pale yellow-green was the desired colour here), to see if we were putting in the right dose. If the chloride levels were too high then we would blow-down the boiler as detailed previously; this would of course get rid of the chemicals we wanted as well as the ones we didn't, so after the retest we would invariably have to redose as well. All this testing, dosing and blowing down would usually occupy us for the first two hours of the midnight to 4 am watch, following which we would return to the control room for a well-earned cuppa.

After a few days of being on watch with GW it became obvious that he trusted us (me and the junior that is) to do all this boiler treatment work by ourselves, which we were happy enough to do, as I had found that the night watches were always pretty boring. GW himself must have thought that with us in charge of this important work he could afford to let himself go a bit, and used to spend every evening drinking in the bar and seldom went to bed much before 2200, which only gave him a couple of hours to sleep it off. The result of this was that he was usually still drunk as a skunk when he came down on watch at midnight. After a particularly heavy session he would sit miserably in the control room for the first hour or so of every watch, drinking coffee and being sick into the rubbish bin. Sometimes these drinking spells would last three or four days, during which he seemed to be getting ever thinner and weaker, but eventually he would manage to dry himself out a bit – probably after being carpeted by the chief. The ultimate sanction would be for him to have his tap stopped i.e. being refused any bond issues. I dare say that if this had ever happened he might have died of malnutrition!

As the voyage progressed I soon settled into the routine of the 12 to 4 watch, although, for the reasons given previously, I never liked it much. Once a week we would have lifeboat drill in the morning – usually after breakfast, which would spoil our morning sleep on that day (the 12 to 4 never got up for breakfast). Board of Trade sports, as this exercise was commonly called, involved the ship's whistle being given six short blasts followed by one long one, to summon us all to the boat deck with our lifejackets; only the engine room and bridge watch keepers were exempt from this performance. Once at their station, the seamen would undo the gripes, swing the boats out and lower them level with the deck. After this, a couple of seamen, one of the deck officers and an engineer would climb into the boat, the engineer would start the engine and run it for a few minutes to warm it up, following which the signal was given that the exercise was over, and the boats would be hoisted back up and re-stowed for another week.

Board of Trade sports notwithstanding, I found I could never sleep in much beyond 10 in the morning anyway, which meant I would have an hour or so to spare before lunch. If the weather was half-decent I often used to go for a stroll down to the forecastle to stretch the legs and get a breath of really fresh air. Standing right up there in the bows, some 200 yards away from the noise and vibration down aft, you could almost forget you were on a ship at all. If the sea was really calm, the bulbous bow could be seen below the surface, slicing through the limpid blue-green water, and once or twice I had the good fortune to see dolphins as they played around in front of us, surfing in and out of the bow wave and effortlessly keeping up with the ship. On the other hand, if the weather was bad, I would go up to the bridge and watch as the waves smashed over the focs'l head. As the spray leapt over the rails to either side, it seemed to be caught in some vortex in the air, which transformed it into a pair of curling wraiths which would drift back toward the bridge windows, appearing to accelerate as they got closer, until they rattled against the glass and obscured our vision for a few seconds.

There is only one other incident that I can recall on our way out to the Gulf, and this happened during a film show one evening, as we entered the Mozambique

Channel. We used to show the films in the officers' saloon, with the projector sited in the adjoining pantry, set up on a stand to project through the serving hatch. If the weather was at all rough, the stand would be tied up to prevent it from falling over if the ship rolled too much, but on this occasion the sea was almost flat calm and we hadn't bothered. All went well until we were halfway through the second reel when the ship suddenly rolled slowly but heavily over to starboard. The picture on the screen seemed to slip sideways across the bulkhead with everybody's heads swivelling round to follow until it disappeared from view and about one second later, a loud crash from the pantry let us know that the projector had fallen off the stand. Amazingly, no serious damage had been done, apart from a dent in the casing and the take-up reel having come off the spool, so we were back in business again in ten minutes. This time we made sure the stand was securely tied up, but no more big waves followed and the sea remained almost flat. One often reads of freak waves in novels and the popular press: well, I can assure you that they do exist and in recent years satellite imaging has shown that they are in fact a lot more common than was previously thought, the most plausible theory for their existence being some seismic shift of the sea floor, possibly hundreds of miles away.

The Gulf in winter was a lot pleasanter than on my previous trip, and the engine room temperature barely got past 100°F on the plates – chicken feed! Our destination this time was Ras Tanura, one of the main oil exporting terminals of Saudi Arabia. If I had thought that Kharg Island was a grim place, Ras Tanura was even worse. Although there was a small port on the mainland, all the big tankers tied up at artificial steel islands a few miles offshore to load, so there was no point in even going down the gangplank to stretch our legs. We were fully loaded in about 24 hours and on our way back again, with the prospect of another month at sea ahead of us – not much fun really, but the money was good.

I think that we were about three days into the return voyage when the first signs of condenser trouble manifested themselves. I had been doing the boiler tests that night and found that the chlorides on both boilers had jumped by 14 ppm in 24 hours. I duly reported this to GW, who went and did a test for himself, which gave much the same results as mine. A check on the salinometer, meanwhile, showed it to still be in the green sector but noticeably nearer the yellow than on the previous watches; this indicated that one or more of the condenser tubes had started to leak. GW decided that we would blow down and dose as per usual on this occasion, rather than wake John up straight away to give him the bad news. In the meantime we would attempt to seal up the leaking tube using sawdust which was a new one on me.

Seeing my blank expression, he explained that if a large quantity of sawdust is introduced into the main seawater cooling pipe to the condenser, there was a chance that a particle might find its way to the hole and stick there, held in place by the pump pressure on one side and the condenser vacuum trying to pull it through on the other. As wood tends to swell as it becomes saturated with water, this would also help it remain in place, plugging the hole. Next, as the main circulating pump had two strainers on the suction side, each with its own set of valves, then if one of those strainers were to get blocked it would be possible to isolate and clean it while using the other. This arrangement meant the pump could be kept running and an engine stoppage avoided while cleaning was in progress. The procedure then was rather like dosing the boilers with chemical but on a much larger scale: first, we would isolate and drain one of the strainers, then remove the lid and tip in a few buckets full of sawdust before replacing the lid. Lastly we would simply open the valves on that strainer again and close the other pair, which meant that all the sawdust would now be drawn through the pump and sent on through the condenser and finally back overboard. The main circulating pump, as I have explained earlier, was a massive one, so there was no danger of the impeller getting clogged up in the process.

Considering that the main circ' propelled about a ton of water through the condenser every second and that there were over 2,000 tubes, I have to confess that I did not think it likely that a few buckets of sawdust would have much chance of getting to the exact spot on the particular tube that was leaking. Much to my surprise, however, after the second dosing session the salinometer reading dropped back to normal.

When we were relieved at 4 am, we informed the oncoming watchkeepers of the state of play and said it might be advisable to start logging the salinometer reading every hour as we did not expect our sawdust repair would be the end of the matter. In fact it was a couple of days before the needle once again started to slide the wrong way across the scale – and this time the sawdust treatment didn't seem to work. John decided therefore that we would have to stop and fix it properly, and phoned the chief and the bridge to give them the bad news. As we were at least 100 miles from the nearest land at this time, the bridge said we could stop any time we liked, so after ensuring we had all the materials ready for the job we started winding in the throttle and as soon as the shaft had stopped we put in the turning gear and got started.

Before commencing the real work, we had to flash up the auxiliary condenser and its various pumps and air ejector, after which we could transfer all the various exhausts and drains from the main condenser over to it. This done, we could then isolate and shut down the main condenser. All this took about an hour. Once the main circulating pump had been stopped, the seawater inlet and overboard valves could be shut and the remaining seawater drained down into the bilge. With this done the many bolts securing the great end doors of the condenser were removed and the doors swung back on their hinges so we could get inside. The water boxes at either end were so big that in order to get at the upper reaches of the tube banks we needed to rig planks inside to stand on. There was quite a bit of marine life stuck in the ends of some of the tubes, and it smelt very much like the quay in a fishing port when the boats have just come in.

The basic procedure for finding a condenser tube leak was to take a bucket of water and mix in some fluoresceine dye, which came as a brown powder but turned a yellow-green colour, a bit like car antifreeze, when mixed up. The next step was to fill up the condenser with dyed fresh water so that the tubes were submerged. There was an access hatch on the LP turbine casing in the exhaust belt which led directly into the condenser, through which we could insert a hose to do the filling, adding the dye in stages so that it was well distributed. I can't remember how much water it took to actually fill the condenser up but it was certainly many tons.

Having done the filling, the next stage was to inspect the tubeplate at each end with an ultraviolet lamp, which would make any leak show up a vivid yellow-green due to the dye. Having identified the leaking tubes – and I think we found three at this first go – we then hammered in tapered brass plugs to block them off. Because of the great number of tubes, extreme care was needed to ensure that we plugged the same tube at both ends of the condenser. The man at one end would shout 'top rear bank, 7th row down and 12 in from forward end' or whatever was required – the man in the water box at the other end (about 15 feet away) could hear quite clearly anything that was shouted down the tubes. Having plugged the leaky tubes, we would wash the tubeplate off with fresh water to rinse away any remaining dye and make another examination with the lamp. If all was still good, we then drained the condenser down into the bilge to get rid of the dyed water and boxed the whole lot up again. I think the whole procedure took around eight hours before we were on our way once again.

Now that we had done the job properly with plugs instead of sawdust, we all thought that this would be the end of the problem. None of us could have foreseen what a gruelling trip lay ahead for us. Indeed, not even two days passed before our optimism was shattered, as the salinometer reading started to increase once again. Sawdust gave us a reprieve for about another half-day – but then the reading rose rapidly and we were forced to stop at sea for a second time to plug tubes.

I think when it happened this time we were just south of the Comoros Islands, and we had fine weather and a flat calm sea of the most brilliant cobalt blue once again. I had spent an hour or so lazing on deck before I was due on watch and strolled down to the poop, from where I could see a number of seabirds fighting over scraps from the galley that had been thrown overboard. I stood by the rail, idly drinking a beer and enjoying the sunshine when I spotted a dark shape glide past us a few feet below the surface. A minute or so later it was back again, and this time I could see quite clearly that it was a large shark, 9 or 10 feet in length with a white tip to its dorsal fin. With my attention now firmly fixed on the water, which was very clear, I could see that there were in fact at least six of them. I called over to the cook, who was standing by the open door to the galley having a fag (smoking was permitted on the after deck provided we were not working cargo or tank washing), and he said that this was quite normal in tropical waters and that sharks would always find a ship within an hour or two of it stopping at sea. Any thoughts I might have had that we could have lowered a ladder down the side and enjoyed a proper swim were quickly dispelled.

Going down on watch at noon I found the condenser opened up once again and the examination with the ultraviolet lamps in progress. I think we found about five leaking tubes this time, which did not bode well – and also used up most of our stock of ready-made brass plugs. John said what we were all thinking by now: this was not likely to be the end of the matter. He immediately put me to work on the lathe, turning up more plugs from our stock of brass rod.

Sure enough, we had to stop one more time before Cape Town, and on this occasion it was around a dozen tubes that had to be plugged. The weather was not nearly so kind to us this time, and after we had stopped, the ship slowly drifted around until she was beam-on to some quite heavy swells that set her rolling unpleasantly. This made life difficult for us down below, as all our gear kept sliding around over the plates, and the heavy condenser doors had to be tied back to avoid them swinging around and crushing someone.

Cape Town for mail and fresh supplies came and went, and then we made it around the corner and into the South Atlantic before we had to stop again. We were now getting into a routine with it: first try sawdust, which sometimes worked for a few hours and sometimes not at all. Then stop and change over to the auxiliary plant and do the tube-plugging job – and the numbers of leaking tubes were increasing every time. Whichever watchkeepers were on duty at the time would stay down below for an extra two hours after their watch, to assist John and the daywork third with the repairs. On the plus side, we had got the total time taken for the job down to around six hours.

John had been checking the records for the previous refit, which had taken place about two years before, and had found that the condenser had been re-tubed completely at that time. It appeared that an inferior grade of material had been used, which was the reason for the continuing failures. To add to our woes we were starting to run short of all the supplies we needed – sawdust, fluoresceine dye and the ultraviolet lamps, which were very fragile and didn't seem to last very long. I had by now made around 150 brass plugs, which had exhausted all our stock of suitably sized brass rod, so from then on I was making wooden ones by turning down broom handles which seemed to work just as well – I think that I made about 1,000 plugs in total during the voyage. Radio messages were flashed to head office to inform them of the situation, and it was arranged that we should rendezvous with a southbound Esso tanker to scrounge some spares from them.

After yet another stop we finally met up with our fellow tanker and lowered a lifeboat to pick up the spares from it. There was quite a heavy swell running and we had considerable difficulty in recovering the boat when it returned. The pulley block and hooks on the end of the lowering cables probably weighed around a hundredweight (50 kilos) each, and the seamen in the boat had the unpleasant task of trying to grab them as the boat bobbed around underneath, and couple them onto the lifting eyes on the end of the chains attached to the boat – and of course both ends had to be attached before we could commence hoisting. Luckily this tricky feat was achieved without anyone having their hand smashed or getting knocked overboard by a swinging block, and we could finally start the recovery winch. With the boat rising and falling some 10 to 12 feet, the cables were alternately going slack as the boat was lifted and then pulling up tight with a bang as the wave dropped away and it must have been a most uncomfortable experience for the boat crew until they were finally clear of the water. Lifeboats of course are designed first and foremost to be easy to lower away – recovery is very much a secondary consideration, and in fact it was many more years before I next saw a lifeboat actually in the water.

With about another ten sacks of sawdust, four ultraviolet lamps, a fresh tin of fluoresceine, some more boiler treatment chemicals and about a dozen broom handles, we felt that we stood at least a sporting chance of making it back to Fawley – and so it proved, although we had to stop another six times before we got there. Our final stoppage was in the Bay of Biscay which lived up to its evil reputation for bad weather; the ship rolled stiffly and uncomfortably the whole time – I think that on this last occasion we had to plug over 50 tubes!

When we arrived back at Fawley, quite a few of the top brass from the office came down to discuss our problems, and they arranged for the ship to be sent off for an emergency refit, to get the condenser completely re-tubed once again. The captain had requested that we all put on our No. 1 uniforms for dinner after our arrival – presumably so that the bigwigs wouldn't see what a scruffy lot us tanker men really were – and it turned out to be quite a grand occasion, with the cook excelling himself and putting on a superb meal.

Nobody was making much in the way of small talk until the dessert course was being served: one of the dishes was rice pudding, which for some reason had apricot jam in the middle instead of the usual raspberry. The fourth engineer, who was a Lancastrian with a very broad accent, took one look at this, called the steward back and said to him in a loud voice 'Hey-oop lad, what's this, then? – looks like someone's put a gert big dollop o' bastard marmalade int' me rice pudding!' This remark caused everyone within earshot to fall about laughing, after which dinner was a much happier affair. I seem to recall the bosses complimenting the engineers on being able to bring the ship and her multi-million-dollar cargo back safely and less than three days late, despite having to stop at sea 11 times, which they thought was a good effort – and so it ruddy well was!

Esso Warwickshire, which had been built in 1962, lasted until 1988 before being scrapped, whereas *Yorkshire*, a year younger (and in my view a much better ship) was broken up in 1975 – a very short life. Why this should have been I don't know, although by the mid-seventies the availability of diesel engines of 25,000 plus horsepower, together with a sharp rise in the cost of oil, thanks to the sheikhs realising that they had the West over a barrel (an oil barrel in fact!), meant that the writing was on the wall for steamships anyway, as diesel is very much more fuel-efficient than steam: more's the pity.

As I had by now completed my five-month stint, albeit on two different ships, I was not sent to refit and went home to enjoy a month's leave instead.

7 FIVE MONTHS WITH AN 'UP AND DOWNER'

After my leave, which passed all too quickly, my next ship was *Esso Preston*, a coastal tanker of 2,790 tons deadweight. She had been built for the bitumen trade, and all her cargo tanks had been fitted with steam-heating coils to prevent the bitumen from solidifying. For me, however, her most interesting feature was that she was powered by a triple-expansion steam reciprocating engine of around 1,350 hp. I had always thought I would be too late to serve on 'up and downers', which were almost dead and buried by 1968, so when I heard of my posting that March I was thrilled to bits.

I was instructed to join the ship at Purfleet, a port on the Essex side of the Thames between West Thurrock and Rainham – a particularly dismal and depressing part of the coast, especially back in those days before the modern Lakeside development had been built in an attempt to inject some prosperity into the area. My parents drove me up via the Blackwall Tunnel, and it was late evening when we eventually found the ship, having groped around the unlit streets beforehand for some time while looking for the entrance into the terminal. The tide was right out and the jetty was about on a level with the bridge of the little ship, which meant a very steep climb down the gangway to the deck.

The ship was almost deserted apart from a seaman tending the ropes and the second mate who was keeping ship. Quite what the crew had managed to find in Purfleet in the way of entertainment, or even a decent pub, I have no idea. The second mate had at least been expecting me, and directed me to a quite spacious cabin in the bridge superstructure; all the officers including the engineers were quartered amidships, while the crew lived down aft. I found a pantry where I made my mum and dad a cup of tea before they had to make their way back home again.

I wandered down aft and poked my nose into the engine room where I was greeted by my first view of the triple-expansion engine, which was a real beauty, and the two Scotch boilers that supplied the steam for it. The boilers were still fired up and were not far off their full pressure of 225 psi, although none of the other steam plant was running and all the electrical requirements of the ship were being met by a 6-cylinder diesel generator – I think it may have had a Perkins engine but after all this time I can't be sure. When I had got down the ladder to the bottom plates I was met by the third engineer, who greeted me and gave me a quick guided tour.

The engineering crew of *Esso Preston* consisted of the chief engineer – an alcoholic Geordie approaching retirement, who never once to my knowledge set foot in the engine room, a second engineer, the third engineer, as mentioned above, and a fourth, together with one fireman/greaser on each watch. The second engineer was not in fact a proper engineer at all and had started his career at sea as a cabin steward with the mighty Cunard Line. He had a foul mouth and an even worse temper, so I can't imagine how he had managed to hold down a job as a steward on a passenger ship. He had been on *Preston* for so many years working in the engine room, however, first as a rating and then by virtue of long experience, as fourth and then third engineer, that he had been granted a dispensation by the Board of Trade to sail as second without actually having to take his second's ticket. I am quite certain that this dispensation would only have been valid on coastal trade vessels like *Preston* – and possibly only on the *Preston* herself, as he never mentioned to me that he had ever been on any other of the Company ships. I think it rankled with him that when the old chief retired it would be the third (who did have a proper ticket) that would take over the job, rather than him.

After the complexities of turbine ships, the dear old *Preston* was wonderfully simple, and within a few days I had worked out all the pipe systems and could pretty much run the job on my own. The second obviously thought so too, because he put me on his own watch (the 4 to 8) and unless we were on stand-by while entering or leaving harbour I never saw him in the engine room before about 7.30 in the mornings. This didn't bother me at all, and in fact the less I saw of him at these times the better I liked it, as he too had a drink problem and was usually nursing a hangover every morning, which made him all too ready to work off some of his bad temper on me.

As to the machinery, nearly everything was steam-driven – quite a bit directly from the main engine itself – and apart from the lights, galley equipment and the domestic fresh water and sanitary water pumps, there was hardly any other electrical plant. The two Scotch boilers had small Weirs-type fuel and feed pumps, while the forced-draught fans were driven by a couple of high-speed enclosed crankcase steam engines (and if they broke down or when first starting up, the boilers could also be fired just with natural draught if the need arose). The seawater cooling pump for the condenser, a bilge pump, and the Edwards air pump for creating the condenser vacuum were all driven from the main engine high-pressure crosshead. The steering consisted of an old-fashioned steam tiller which was normally motionless apart from a wisp of steam leaking from the glands to tell you it was actually turned on; when it got a helm order, however, it would burst into life like some demented robot and clatter over to the required position before subsiding into brooding silence once again. For our limited electrical needs there was a very small steam turbo generator to supply the load at sea, and the aforementioned diesel to do the same when we were in port.

The triple-expansion engine stood around 10 feet high from the floor plates to cylinder head level and was kept in a very clean condition, with all the copper and brasswork being polished at least twice a week, while the steelwork was kept bright and shining with Scotch-Brite. The second engineer was largely responsible for this, and was forever going round wiping the engine over with an oily rag or attacking any rust spots with emery cloth before they had a chance to spread. Whatever his other faults, the engine was always a picture and did him much credit. The chequer plating around the engine was also kept in gleaming condition and was mopped over every watch. At the foot of the ladder leading down from the top level, the usual route in and out of the engine room, legions of engineers' boots had worn away the diamond pattern, so that it was now just a smooth sheet of bright steel plate.

The triple-expansion engine had been invented well back in the 19th century and is so named because although the engine may have had three cylinders, the steam from the boiler is admitted only to the first of them – the high-pressure, or HP, cylinder. After doing its work there, it is exhausted into a second, larger, cylinder called the intermediate-pressure, or IP, and the exhaust from this goes in similar fashion to the low-pressure or LP, which is bigger still. Finally, the exhaust from the LP goes to the condenser, where it is turned back into water, so the steam is used (expanded) three times.

As these engines go the one on *Preston* was a quite modest affair; I seem to recall that the HP cylinder was 18" diameter, the IP was 28" and the LP around 42", with a common piston stroke of 30". As you can see, even an engine this size had a pretty big LP cylinder, and some of the very large triples used two LPs, to avoid the manufacturing difficulties of constructing cylinder blocks with bores of 8 feet or more in diameter – although the latter would now have had four cylinders, they would still be triple-expansion. The engines of *Titanic*, amongst the largest ever built, had cylinders of 54" and 84", and two LPs at 97" diameter with a piston stroke of 75". Those two giant 'up and downers' weighed around 1,000 tons each, and together with the boilers will still be largely intact on the bottom of the Atlantic Ocean long after the rest of that great ship has turned into a pile of rust.

Unlike my previous deep sea ships, *Preston* was engaged on coastal and near continental trade only, and we were therefore in and out of port every few days. I can remember that our first trip from Purfleet was to Antwerp, where the third engineer took me ashore. We went to a place called Danny's Bar, where there were a number of extremely good-looking women to be seen mingling with the customers. I have to confess that I was very attracted to one of them and mentioned this to the third, who burst out laughing. 'I should be careful if I were you,' he said, 'I think you may find that woman is in fact a bloke in drag!' I found this pretty hard to believe but when 'she' sidled up to me and asked if I wanted to buy her a drink in a distinctly baritone voice, I finally saw the light (and also a trace of five o'clock shadow under her make-up.) Having survived this embarrassment, we enjoyed a few beers and listened to a local group, so it finished up as a very pleasant evening.

From Antwerp, it was back to Purfleet, then to France and a trip up the River Seine, where our destination was a tiny place called Quillebeuf-sur-Seine. Here we found a typical French bar, where we sampled some of the local wine instead of beer – I was not impressed, although we also had a meal which was both tasty and cheap.

The following morning I was down on watch by myself as usual, when the phone rang: this was the bridge, who enquired 'Could we be ready to go in three-quarters of an hour?' Provisionally we had been expecting to sail after breakfast, so this left me in a bit of a dilemma as to whether or not I should go and wake up the second engineer for stand-by. I had never actually prepared the engine for sea by myself, but on the other hand, the second had been drinking with the chief until well past midnight. This meant he would not be in a very fit state to do much anyway, and could be pretty well guaranteed to give me a mouthful of abuse if I woke him up, so I decided to have a go myself. The fireman on my watch was quite a useful sort of chap and could look after the boilers on his own, which meant that all I had to worry about was getting the engine ready, putting steam on the steering and running up the turbo genny.

I attended to the genny first, which, as *Preston* was a DC ship, meant that there was none of the synchronisation procedure to go through that is associated with AC plant; it was simply a case of running up the turbine until the governor took over, ensuring that it was actually producing some voltage, and then throwing in a knife switch on the very old-fashioned DC switchboard. The load would then be balanced between the two, using the trimmers on the field coil regulators. When we were Full Away at sea, the diesel genny could be disconnected by throwing out another knife switch and then shut down.

Warming through the engine consisted simply of opening all the cylinder drains so that any condensed water in the cylinders could escape, starting the reversing engine so that it went from ahead to astern and back again about every 15 seconds, and then cracking open the throttle so that between about 10 and 20 lbs of steam showed on the high-pressure steam chest gauge. At first, nothing would happen except for the sound of steam and water blowing out of the drains, while the weighshaft which operated the Stephenson valve gear would move slowly to and fro, taking the expansion links and eccentric rods from side to side with it. After a minute or two of this, the engine would give a convulsive movement one way or the other and then, as the reversing gear moved over to the opposite direction, it would kick back again; steam reciprocating engines run equally well either way and are reversed by resetting the valve gear using the reversing engine as described above. These oscillations would get steadily larger as the engine warmed up, until eventually it was making a complete turn or so each way – I never ceased to be fascinated by the sight of all the gleaming steel rods and cranks slowly coming to life.

The engine could be left on its own to kick over like this for several minutes, but as it warmed up it would steadily become more energetic until it might be making a couple of full turns in each direction; at this stage we would close the throttle in a bit because we didn't want to be turning the propeller enough to start moving the ship and straining the ropes. At around the same time the HP and IP drains could also be closed, as by then they would be blowing mostly steam, although the LP ones were usually left open until we were actually on the move – the LP cylinder being the largest, took longest to warm up.

The reversing engine was a little two-cylinder vertical job sited conveniently beside the main throttle, so that both were within easy reach of the man on manoeuvring duty. It was operated by a small valve on the steam supply which was of the quick-acting variety, so that it could be started and stopped easily and precisely where required. The engine drove a worm and worm wheel, to which was connected a crank pin and rod that operated the weighshaft as noted above. On the worm wheel, which was of burnished steel, two brass plates, marked Ahead and Astern, were riveted on, 180 degrees apart. To manoeuvre the engine, it was necessary to start the reversing engine running until the appropriate brass plate lined up with a steel pointer, and then stop it, at which position the engine was set to run in the direction indicated.

Unlike turbine plant, reciprocating engines could be reversed very quickly, no more than a few seconds being required to go from ahead to astern. This rapid manoeuvring capability made the steam reciprocating engine the popular choice for tug boats and small ferries up until the 1960s, when controllable-pitch propellers became commonplace, thereby allowing the all-conquering diesel engine to take over this last bastion of the up and downer.

While the engine was warming itself up, the other main job I had to attend to was the lubrication. Unlike turbine ships, where oil is pressure-fed to all the bearings, triple-expansion steam engines are lubricated by hand in the same way as railway locomotives or traction engines. There was a small mechanical pump that provided minute amounts of thick steam oil to the high-pressure cylinder – this used about two pints every watch. The more important bearings were supplied by brass oil boxes with copper pipes leading down to wherever the oil was required. Each of these copper pipes terminated inside its box just below the top, and the oil was siphoned down them by worsted wicks twisted into pieces of soft iron wire – the more strands of worsted there were in these trimmings the faster the oil would siphon out. Every time we were getting ready to put to sea, the boxes would be filled and the trimmings inserted into the pipes to start the oil flow; at Finished with Engines they would be taken out again to save wasting oil. All the lesser bearings and pin joints were supplied from oil cups or even simple oil holes, which would get a squirt from an oil can once or twice a watch. I was amazed that this large engine would happily run all day on just a couple of gallons of lubricating oil applied in this rather haphazard fashion without any bearing ever running hot – at least none ever did when I was there.

Just before the appointed time for departure, the bridge phoned down to ask if I was ready, and having received my affirmative answer, tested the telegraph by ringing round to each position in turn before coming to rest on Stand-by. It wasn't a very long wait before I got the first movement:: Slow Astern. I started the reversing engine running and then neatly stopped it when the appropriate brass plate was lined up with the pointer, before giving the main throttle a quick half-turn, which caused the engine to ease smoothly and silently into motion. There was no rev counter to tell me how fast the engine was going, but the HP steam chest pressure gauge was marked in red at the pressures that corresponded to the required speeds, so all I had to do was to adjust the throttle until I got the right pressure. The engine was really delightful to handle, and although the movements were coming down every few seconds until we cleared the berth I was managing to keep up with them quite easily; in fact the hardest part of the job was finding the time to record them all in the movement book, which finished up looking a bit untidy as a result.

Occasionally the engine might stop with the HP piston on top or bottom dead centre, from which position it could develop zero torque, so the engine would not start again. To get round this problem, the IP cylinder (whose crank was set at 120 degrees round from the HP) could be given steam directly via an additional control called the simpling valve, which bypassed the HP cylinder; one puff of steam applied in this way would be enough to start the engine moving, after which the main throttle could be opened in the normal way.

After around ten minutes of shunting back and forth the telegraph settled on Full Ahead, and I guessed that we had swung round and were now on our way back down river. The fireman meanwhile had been busy adjusting the fuel and feed pumps, and had by now got all three furnaces lit up in each boiler, so we were generally in good order and gave each other an encouraging thumbs up.

It was now getting on for 8 am, and the friendly figure of the third engineer came down the ladders to take over. Normally, the third would have had the 12 to 4 watch, but on *Preston* he had swapped with the fourth who for some unknown reason didn't seem to mind. Seeing that I was on my own he remarked: 'It didn't take that idle bastard long to get you trained up, then?' referring of course to the second. I replied that he was usually in such a foul mood every morning that I was a lot happier that way. We had a bit of a chat about him, which was when I learnt the bit about him having been a Cunard steward. The second himself finally appeared just after 8 am, looking rather bleary-eyed. I would imagine that most people might think that a word of thanks would be in order, seeing as I had covered his job for the entire watch: they would be mistaken, however, for he walked past us both without a word – not even 'Good morning'. He then spent a minute or two examining the log before coming up to me and remarking what a mess I'd made in the movement book and why hadn't I swabbed the plates?

The third, on the other hand, was a really decent type and had actually got a chief's ticket; this would normally have guaranteed him a job straight away as a second engineer on one of the company's deep sea ships, but I think that he had some ongoing domestic situation and had requested to stay on the coastal ships so that he could quickly get home if the need arose. We got on really well, and whenever we were in Fawley, which was quite close to Southampton, where he lived, I would sometimes stand his evening watch for him, so he had a chance to go home for the night. He in turn would do one for me in other ports, so that I could get a decent run ashore. As the second hardly ever bothered doing a port watch anyway, this meant I was sometimes in full charge from 4 pm until 8 am the following day, although it has to be said that the dear old *Preston* did not require a great deal of attention in port, so I could spend a fair bit of the time in the mess, drinking tea and reading.

On one occasion when we were in Fawley I fancied a trip into Southampton to see if I could find a music shop to get another set of strings for my banjo, which I was attempting to learn at the time – much to the dismay of all those in the midships accommodation. The driver of the company minibus also had business in the town centre and took me all the way there straight after breakfast. This was a stroke of luck, otherwise the lift would only have been as far as Fawley refinery gate. I wandered around the town for a while until I found what I was looking for, after which it was the usual pub for lunch.

Still having an hour or two to spare before going back to the ship, I strolled around for a while and found myself in a park where a most magnificent memorial to the engineer and electrical officers of *Titanic* had been erected – all 35 of whom had gone down with their ship. One of the features of that disaster is that many of the survivors reported that the lights remained burning right up to the last few minutes before the ship sank, which meant that all those still left inside the vessel could at least see their way, hopefully, out onto an open deck. The men in the engine and boiler rooms who were responsible for keeping those lights going would have known that the ship was doomed for at least an hour beforehand, and yet they stayed down below trying to pump out the water and tending the generators as the angle of the decks beneath their feet increased relentlessly. Joseph Bell, the chief engineer, had in fact relieved all his men of their duties some 20 minutes before the final plunge, and yet they opted to remain with him at their posts until it was too late for them to escape. This was an example of unflinching courage that is surely hard to match, and I mentally doffed my cap to them all.

One other permanent sign of the tragedy is that following the disaster it became accepted that engineer officers wear purple as part of their rank insignia. It is a myth, however, that King George V was so moved when he heard of their courage that he decreed it to be an official mark of recognition – in fact the Royal Navy had introduced 'branch' colours from around 1860 onwards, and the Merchant Navy, as it had come to be known, was simply following on. Even so, it is a good story, and it means that if you see a picture today of a chief engineer you will know he is the chief and not the captain because the four gold bars on his sleeves will have those purple bands between them.

Although I had soon got used to the watchkeeping routine on *Preston* and the frequent stand-bys, when I was off watch and trying to sleep it was another story. One soon becomes accustomed to the usual shipboard noises and the vibration of the engines, but an intermittent rattle or the squeak of an unsecured door swinging back and forth on its hinges can be extremely annoying and will completely preclude sleep. Eventually you will be forced into getting up and fixing it, and most of the cabins I occupied during my career at sea had pieces of cardboard or beer mats jammed into the cracks around any loose panels to stop them vibrating or rattling.

On *Preston*, though, it was a different problem: in the space above the deckhead in my cabin there was something loose that rolled from side to side as the ship herself rolled, and I couldn't see any easy way of getting in there to remove it – it used to drive me insane. Occasionally in calm water it would stop for a few minutes but I would still lie there awake, waiting for the slightly bigger roll which would start it off again. Eventually I became so paranoid that I began to wonder if perhaps a previous occupant of the cabin with a sadistic sense of humour might have put something in there deliberately in order to annoy the next one.

Finally after about my third sleepless night, I decided that enough was enough and that the deckhead would have to come down, even if I mangled it in the process. So in the morning after my watch I started out on it, armed with a selection of screwdrivers, chisels and a great deal of bad temper. It took a couple of hours to remove the self-tapping screws, as they were all rusted in solid – some sheared off and a few had to be chiselled off – but eventually I got a couple of panels down, which was enough to stick my head through and take a look. At first I couldn't see what was causing the noise but just then the ship made a slightly bigger roll and the culprit, a short length of discarded steel pipe left over from some previous plumbing job, obligingly rolled across right in front of me and I was able to grab it. Oh boy, did I feel better when I slung the little beast over the side as far as it would go!

For the first couple of months of my trip on *Preston*, we seemed to spend most of our time running between Purfleet, Fawley, Antwerp, Rotterdam, Quillebeuf and Falmouth, but after one trip to Antwerp, where we had loaded bitumen, we got orders for Dublin, which we thought would make a pleasant change. Bitumen remains as thick as treacle even when heated, so we expected to have a day or two in port because it was very slow to pump in and out – unlike fuel oil, which we occasionally carried instead, and which was a lot quicker.

The weather on our way down Channel and up into the Irish Sea was quite rough, and *Preston* at times seemed to be struggling to make any progress – her top speed was only around 10 knots. Rounding the corner at Land's End we were beam-on to some quite heavy Atlantic rollers, and the little ship was rolling most uncomfortably for several hours. I was off watch and found that the only way I could stay in my bunk and get some sleep was to stuff my lifejacket under the mattress on the outside, so that I could wedge myself into the angle it made against the bulkhead.

We arrived in Dublin in mid-morning and found our berth was quite close to the city centre, which made a pleasant change from an oil terminal in the middle of nowhere, which was our usual lot. The third was going to do my afternoon watch, so I had the rest of the day to explore the city. There was one other cadet on board, so I waited until he came off watch at 12 and we went ashore together. It was a lovely sunny spring day and we strolled down Custom House Quay and into O'Connell Street, right in the town centre. We spent a minute or two admiring the imposing and famous post office, the site of the Easter Rising in 1916 – some of the bullet marks could still be seen. After this we went looking for a traditional Irish pub and soon found one. This was where I sampled proper draught Guinness for the first time – it certainly made a pleasant change from cans of Tennent's lager. From here we did a bit more sightseeing and got as far as Phoenix Park, where we spent the rest of the afternoon. After that we strolled back toward the town centre and found another pub with a live folk band – singing mostly anti-British songs, I noticed. Everyone we met there was friendly enough, however, and all in all it was a most enjoyable day.

Back on board ship again that evening I learnt that our next port was to be Eastham, at the western end of the Manchester Ship Canal, after which we were going to Grangemouth on the Firth of Forth. This would mean a trip around the northern coast of Scotland and through the Pentland Firth. If we continued from there back to Purfleet, which was our chief officer's best guess, it would mean we would have completed a circumnavigation of the British mainland.

Our trip across the Irish Sea on the first leg of this voyage was enlivened by the air release cock of the engine driven condenser cooling pump, which suddenly blew out of the top cover. This meant that at every stroke of the engine (and it was running at 120 rpm at the time) a solid jet of cold sea water shot out with such force that it reached the deckhead and bounced back down again. In a matter of seconds everything within about 20 feet of it, including us, was soaked. The second was aghast that all his lovely shining steel was now getting water-jetted, and yelled at me to shut the throttle. This done, the engine slowed right down, which lessened the force of the water, but it was a minute or two before it stopped completely, as the wash over the propeller was enough to keep it turning until the ship lost way. Meanwhile, the sudden reduction in the demand for steam had lifted the boiler safety valves, which roared away up the funnel until we were able to shut off most of the burners and reduce the fuel pressure.

We tried to phone the bridge to let them know that we would be stopped for a few minutes, but the phone appeared to have drowned, so the fireman was dispatched topsides to let them know the reason for our abrupt halt. The air release cock itself had disappeared somewhere down into the bilges, so rather than waste time looking for it the second sent me up to the stores to find a tapered steel BSP plug which we fitted instead, and in the end it was only about five minutes before we could get going again. Surprisingly perhaps, the phone and one light fitting were the only items of electrical equipment to have suffered under the deluge, and these were soon repaired.

The rest of the watch was spent mopping up and wiping over the engine (the bits that weren't actually spinning round, that is) with an oily rag to stop them rusting. Just for once I felt sorry for the second engineer; the engine really was his pride and joy and now we could practically see the salt stains appearing before our very eyes over the copper and brasswork as it dried out, while the burnished steel handrails around the top of the engine were showing rust spots even before we went off watch. I guessed that as soon as we were next tied up our arms would be getting plenty of exercise with the Scotch-Brite and Brasso.

During our trip across from Dublin the engine had developed a slight knock from the HP bottom end bearing (on a ship they are always bottom ends, as opposed to big ends in a car) – perhaps it hadn't appreciated being lubricated with sea water during the crossing. At any rate, the third decided he would take it down for inspection on his watch, and invited me to assist if I felt so inclined – and as Eastham didn't look as if it would have much in the way of tourist attractions, I was happy to agree.

First off we needed to put the HP crank on top dead centre to give us room to drop the bearing out underneath. The turning gear on *Preston* was a worm and worm wheel device, and was strictly handomatic. The worm wheel had to be swung into engagement with a toothed wheel on the propeller shaft and was then locked into place with a steel pin. After this a 3-foot steel bar with a ratchet was fitted onto the worm shaft and the engine could be barred round to the right place. It was a slow and arduous process, and it took us about ten minutes to move the engine a half-turn, although in this case the third had neatly stopped the engine pretty well in the right place after the final movement so we didn't have far to go.

With the HP piston on top dead centre, we clamped the piston rod in place so that it could not descend under its own weight once the bearing was removed – I am not sure how we did this now, although a bolt screwed into the guide under the crosshead bearing would have done the trick. This done, we loosened the two bottom end nuts so that the lower half of the bearing began to drop away from the crankpin. With the third now standing in the crankpit and taking the weight, I took the nuts and bolts right off, which allowed him to get hold of it and swing it out onto the plates. We were fortunate that the engine was of a size where this could be done manually – if it had been much bigger we would have needed to rig lifting tackle, and the job would have taken much longer. As the two halves of the bearing split apart we had to be very careful not to lose any of the brass shims between them, which provided the clearance.

To remove the top half of the bearing it was necessary to bar the engine round some more, which left the connecting rod behind while the bearing came round with the shaft – I seem to remember we secured the connecting rod with a rope so that it wouldn't swing back suddenly as the bearing came away. The third remained in the crankpit, keeping hold of the bearing until there was enough clearance to lift it off and dump it on the plates as before.

With the bearings now safely removed we could have a good look at them and decide upon our course of action. The two bearing halves were of steel, each with a bronze insert (called a brass) that was spigoted into place. The actual bearing surfaces consisted of white metal which had been cast into the brasses and then machined out to nearly the correct size before being finished by hand-scraping. The bearing surfaces appeared to be a dull grey colour with a few shinier spots here and there. The third said that the surface looked as if it had work-hardened a bit and he proposed to give it a light scrape all over to clean it up, after which we would check the fit and the clearance. The scraping was done with a three-cornered hand scraper, which soon had the surface looking bright silver again.

Now we had restored the white metal to give a decent bearing surface, we had to check the fit to ensure there was an even contact with the crank pin. To do this we put a very light rubbing of Micrometer Blue over the journal, then put the bearing brasses back on in turn and wiggled them back and forth a few times so that when we removed them the places where the bearing was touching would show a blue marking. This done, it was out with the scraper again to carefully remove the high spots where the blueing was heaviest, and then repeat the process several times until we had a decent contact over most of the bearing surface. Micrometer Blue is like a thick blue grease and comes in small tins, about half the size of the average tin of shoe polish; it is so concentrated that the tiniest amount can be spread over a very large area. The less you use the better, in fact, because too much will simply give you an overly optimistic impression of the bearing fit. This makes it so economical to use that I still have the original tin that I bought 40 years ago.

With the bearings now nicely scraped in, the final part of the job was to check the fit of the bearing on the crankpin – as we had scraped off a fair bit of white metal, the clearance was bound to be too much and would require adjusting. To measure the clearance we stuck three lengths of .025" lead wire around the top of the journal with grease and then carefully laid the top half of the assembled bearing into place over them. Now the third lifted the lower half of the bearing, complete with all the original brass shims, up into place while I quickly put in a nut and bolt to keep it there. When we had both bolts in place we tightened them up hard, which had the effect of squashing the aforesaid lead wire. Now we had to dismantle the bearing again and carefully peel off the lead strips from the journal so that we could measure their thickness with a micrometer.

A useful rule of thumb for a white metal bearing of this type is to allow 1/1,000" clearance per inch of bearing diameter. In our case we had about 10 thou too much, so we simply applied the micrometer again to measure all the bearing shims and took out this amount from each side. The third was very meticulous and insisted we did another check with the lead wire before he was satisfied, but having established that the clearance was in fact now correct, we boxed the job up and went topsides for a well-earned beer.

When we sailed again, the engine was once more running like the proverbial sewing machine, with no sign of any bearing knock. The second had not been idle, either, and had repolished all the bright work, so there was no sign that the whole engine had been drenched in sea water only a couple of days before. I had by now come to the conclusion that cleaning and polishing was about the only useful function he performed on *Preston*, because his engineering skills were very limited, as the following anecdote will show.

On the face of it, the word 'drill' would not appear to invoke much in the way of mirth but following this incident, the third and I could be almost guaranteed to crack up every time we heard it mentioned on board *Preston*. It started off one morning when we were in port and the second asked me to help the third plumb in a new water supply somewhere – I forget the details now but it involved drilling a 1.25" diameter hole through the steel deck. This was far too big for our electric hand drill, and the second gave us a very large and heavy pneumatic one to do the job. I drilled a 3/8" pilot hole first, which apart from making it easier for the big drill to follow, also gives you a more accurately sized hole.

The problem with opening up a pilot hole is that unless the follow-up drill is very carefully ground to the correct rake and clearance angles, it can almost be guaranteed to bite suddenly and jam up. I don't wish to appear boastful but one thing I can do pretty well is to sharpen a drill, and I was happily engaged on the grinder in the workshop touching up the 1.25" drill when the second appeared. He watched me for a few seconds before snatching it out of my hand and saying in his usual charmless manner, 'Give it here, you haven't got an effing clue.' He then proceeded to regrind it again with me and the third watching discreetly; when he had finished it had excessive rake and the angle on the point was far too steep. The third came to my rescue and suggested to the second that having showed us how to sharpen a drill he should also give us a demonstration of how to use it. Of course, he could not now back down without losing face, so we all trooped out on deck to watch the performance.

The drilling machine was a massive tool operated from a compressed air line and had a couple of steel bars sticking out the sides to hold it by, one of which had the operating trigger mounted on it. What happened next could not have been choreographed better by the director of a slapstick film. The second got himself in position, put the point of the drill in the pilot hole and squeezed the trigger. As we had expected, the drill completed less than a half a turn before it grabbed violently and stuck – but the second failed to let go the trigger and as the drill was now unable to move, it was the machine itself which continued to turn round, taking him with it. He had also failed to notice that his foot was placed in a loop of the airline, which tightened up around his ankle, neatly whipping him off his feet and dumping him unceremoniously onto the deck, when he finally remembered to let go of the trigger. He stormed off, effing and blinding furiously, and we didn't see him again until dinner. It took us several minutes to stop laughing enough to regrind the drill again and finish the job.

We had fine weather almost the whole way up the west coast of England and Scotland, and some of the views we had of the Western Isles were quite stunning. The Pentland Firth also treated us kindly, and we passed around the corner at John O'Groats and into the North Sea in glorious sunshine. It was to be nearly 40 years before I was to see the Pentland Firth again, and this time it would be from the landward side, when my son and I had completed a charity bike ride from Land's End to John O'Groats.

At one of our bitumen ports, where we were tied up for a couple of days, the second decided we should do an external inspection of one of the boilers – and by 'external', I don't mean walking round the outside of it to inspect the silver paint on the lagging. This job involved crawling through the furnaces to reach the combustion chambers at the far ends and then checking the tube plates. I can hear you saying: 'So you are actually inside the boiler but it's an external inspection?' That's a fair question., but when we say 'external' on a boiler, we mean the fire side spaces, as opposed to 'internal', which are the water spaces.

The boiler was allowed to cool down for 24 hours, after which we removed the oil burner assemblies to gain access into each of the three furnaces. Then we could crawl through them to the combustion chambers at the back ends. Once there, it was possible to stand up and inspect the tube plates; we were looking for any signs of leakage or cracking of the tube plates between tubes, and checking the tube ends themselves for signs of thinning or burning, which might indicate poor combustion or simply that they were near the end of their life.

The boiler still had quite a bit of steam pressure on and it was extremely hot inside, especially during our crawl through the furnaces where the metal was still much too hot to touch with bare hands – no place to be if you suffer from claustrophobia. What made it bearable was that we could open the funnel damper, which meant there would be some natural draught to pull a bit of fresh air through and save us from being slowly roasted to death. This time all appeared to be in order, so we now turned our attention to the tube plate at the other end: this was much easier as it could be reached from the outside by swinging open some large access doors above the furnace fronts.

When we had finished we replaced all the burners and lit up one of them up again to start raising steam pressure. This was done with a torch made from a steel rod about 3 feet long, around the end of which had been tied some asbestos string. This was dipped in paraffin and lit with a match, after which it could be inserted through a small opening beside the burner and the oil turned on. As soon as the oil spray had ignited, which it often did with a muffled 'whoompf', the air register was opened enough to clear the puff of black smoke that always accompanied the initial firing up.

Water-tube boilers have very large furnace spaces, a relatively small water capacity and extremely good circulation, which means that steam can be raised very quickly. On one of my later ships in an emergency, I once saw the steam pressure raised from nothing to 860 psi in less than half an hour with no apparent ill effects. On a Scotch boiler, however, the furnace spaces are small, the water spaces are large and the circulation is almost non-existent, so steam has to be raised very much more carefully to avoid uneven heating of the metal, which could cause cracking. We now had to spend several hours bringing the pressure up from about 80 psi to the full 225. Job done!

In the same way that I had been shown the basics of astronavigation on *Esso Yorkshire*, on *Preston* I had befriended the second mate, who showed me how to navigate around the coast by taking visual and radar bearings from shore marks and lighthouses etc. After I had finished my evening 4 to 8 watch I would quite often spend an hour or two on the bridge seeing how the other half live. Occasionally (and always under the rather sardonic eye of my friend) I would be allowed to take some bearings and put my X on the chart at the place where I thought we might be, following which he would mark his fix. Now and then we would actually be in agreement, give or take a quarter-mile or so, which made me feel pretty good.

Radar in those days was nowhere near as sophisticated as it is now. If we saw a blip and wanted to know whether it was a ship on a collision course or not, chinagraph pencil marks would be applied directly to the radar screen every minute or two as the unknown vessel got closer, until there were enough to draw a line through the dots which would show the relative bearing: if the line happened to be pointing at the centre of the screen (one's own position), it would mean the vessel was in fact headed straight at us, and avoiding action would or would not be taken according to the maritime rules of the road. These days, a click of a cursor on the blip is enough to acquire a target, and after a few sweeps of the scanner the screen will be able to show the relative vector, true vector, course and speed of the target – and probably, if the march of technology continues at its present pace, the ship's name and what the Old Man had for breakfast!

Grangemouth is on the Firth of Forth quite a bit further west than Edinburgh, which meant we would be passing beneath the two great bridges which span that particular stretch of water. The first one was of course the famous railway bridge which reputedly provides continuous employment for a gang of painters – in fact the expression 'painting the Forth Bridge' has entered the language to mean any job which never seems to be finished. Apart from the painting, this bridge requires little in the way of any other maintenance, and although it was completed in 1890, when locomotives and trains were less than half the weight they are today, it still stands firm. The road bridge, on the other hand, which was opened to traffic with much ceremony in 1964, had by 2003 developed corrosion problems in the main suspension cables. There is no easy fix for this as the cables are made up of many thousands of individual steel wires which are impossible to replace or repair, so that eventually the decision was taken to build a completely new bridge alongside the original one. The new bridge opened in 2017, while the 1964 bridge has been downgraded and no longer carries HGVs. From Grangemouth we went on to Purfleet as predicted by the mate, thereby completing our circumnavigation of Britain. I am not sure after all this time, but I have an idea that we did a similar round trip again before I was finally paid off.

Preston was a really nice little ship, although not exactly speedy – on one passage from the Thames Estuary into the Dover Straits in heavy weather, I can remember seeing North Foreland lighthouse abeam as I went on watch at 4 am – and when I came off again at 8 we had still not got past Dover! On the other hand, what she lacked in speed she made up for in reliability and she was certainly the most dependable vessel I sailed on during my entire career. Unfortunately she went aground off southern Ireland in 1975, and although she was refloated and continued to dry dock under her own steam, it was considered that she was beyond economical repair and finally she was towed to Spain where she was broken up – a sad end for a fine ship.

8 A BLACKOUT AND A PROMOTION

After *Preston*, it was back to turbine ships again, and my next vessel was *Esso Edinburgh*, a tanker of 47,000 dwt built in 1963 by Vickers Armstrong on the Tyne. Unlike my previous ships, this one was constructed with all the accommodation down aft, which apart from the money saved in construction also made the vessel more sociable, as everyone on board was now living in the same block.

Although the ship was still only five years old, she had already had quite a hard life and was by no means as reliable as my two previous big tankers. It's true that *Warwickshire* had broken down with condenser trouble 11 times, but when she was actually under way the job, as on *Yorkshire*, seemed to run itself. *Edinburgh*, however, needed a permanent and close scrutiny of the control panel, and the engineers were forever having to nip off somewhere to adjust something that should have been done automatically but wasn't any more.

Automatic valves controlled just about every variable in the boiler steam and feed systems, and plenty of other things too. They were of the diaphragm type and were pneumatically controlled and operated; the air pressure required was not great – I seem to recall that the supply was set at 25 psi, but the control sensors were very delicate, and to achieve reliable operation the air was supposed to be filtered and dried. The system had been designed by a firm called Bailey Meters, and the main feature consisted of a large panel we called the Bailey board, which was covered in pressure gauges and adjusting knobs. When on stand-by conditions (and quite often at other times as well) an engineer had to be in constant attendance to trim and adjust the settings as required.

It doesn't take a genius to work out that that the Achilles heel of the system was the supply of compressed air itself. On *Edinburgh* we had one small compressor specifically for supplying the control air together with its own filters, dryers and a small air bottle (properly called a receiver). The trouble was that the control air compressor, which had to run 24/7, was pretty well clapped out. Whenever we had to shut it down for repairs, the control air was supplied from the main system via a crossover valve – and this air was neither filtered nor dried, hence the increasing unreliability of the control gear.

On one occasion when we had the control air compressor in bits for overhaul, one of the two main compressors also decided to play up – I think it may have been a bearing running hot. At any rate, we started up the No.2 machine which appeared to be running sweetly enough, so we shut down No.1 and commenced repairs. The gods, however, were obviously not on our side that day, as we had just about got No.1 stripped down when the No.2 gave a loud bang and expired in a great cloud of smoke.

It took about 20 minutes for the air pressure in the main receivers to drop down below 25 psi, during which time we were working frantically to put the No.1 unit back together again. We nearly made it – but this was a race where there was no silver medal for coming second. The automatic valves were either 'air to open' or 'air to close', and they now started to do just that. They were all failsafe, which would mean that most controlled variables would drop to zero rather than go too high. *Edinburgh*'s boilers produced steam at 860 psi for the main engines, alternators and cargo pumps, but everything else used lower pressures – obviously the cook wouldn't want to be cooking his steamed puddings at 860 psi. I think in all there were steam lines operating on at least five lower pressures, each with an automatic valve to control it, or a hand by-pass. In addition to this lot, the water levels in the condenser and boilers, the fuel oil pressure and forced-draught fan pressure were also air-controlled, so in total there were dozens of these automatic valves – which had suddenly all become handomatic.

The trouble was we just didn't have enough men to take hand control of everything, and the job just fell apart: the main vacuum dropped off because there wasn't enough air ejector or gland-sealing steam; the control panel for the boilers had gone to pot so they eventually tripped; and there was a load of other stuff going on too. With the boilers no longer making steam, we had to shut the main engine throttle to try and conserve what we had and keep the genny on line for a bit longer while we continued to run round like the proverbial blue-tailed flies. It was no use. As the steam pressure fell, the genny gradually slowed down and the lights began to burn dim; finally, it too tripped out and we were plunged into darkness. This was my first blackout.

Somebody – I think it was the second – shouted 'Bugger it!' and for about the next 15 seconds we groped around in the darkness. It was eerily quiet with all the engine room fans stopped: apart from a slight hiss of steam escaping somewhere and the whine from the reduction gears as they slowed down, all the usual engine-room racket was stilled. The emergency diesel generator then fired itself up as it was supposed to, and about one in three of the engine-room lights came back on. The second immediately said that there was no point in trying to restart the main plant until we had some air, so the first job was to finish boxing up the No.1 compressor, which we completed within a few minutes. Next we ran it up, filled the air bottles and shut off everything except the vital control airline, which meant we should have enough for at least another half-hour without having to run the compressor again.

Probably in order to save money, the emergency diesel was only big enough to run the emergency lights, a forced-draught fan, a fuel pump and a small boiler-filling pump, which were the minimum requirements for raising steam in a single boiler. There was certainly not enough power to spare to be able to run any engine room ventilation fans, and within minutes we were all sweating buckets and gasping for breath in the hot and humid atmosphere. It also meant that when we wanted to start up the turbo genny and the feed pump (also steam-driven), which would require a condenser and the seawater cooling and condensate pumps just for starters, we would have to stop the boiler again while we did it – how crazy was that?

The idiot who had worked out that we only needed a diesel generator of about a quarter the capacity of the turbos was probably the same man who had decided that it was not worthwhile fitting synchronising gear to it either, so we would be unable to have both the diesel and a turbo generator on the board (connected to the switchboard and supplying power) at the same time. This in turn meant we would have to black out again deliberately while we changed over from diesel to steam. These two factors always made it extremely difficult to restart the main plant.

First we would need to raise steam in the boiler. As mentioned earlier, in water-tube boilers this can be done very quickly, and we got it up from about 300 to the full 860 psi in about 15 minutes. Then it was a case of getting everyone in position to change over from diesel to steam power. The second, with me acting as gopher, stood by at the control panel to restart the pumps and mind the shop; the third was at the turbo genny ready to run that up; the lecky was on the switchboard; one man stayed in the boiler room to relight the furnaces; and the rest were spread round as directed. As for the chief engineer, I'm not sure what he was up to – he hadn't bothered to come down below.

The second gave the word and everyone sprang into action: as soon as we had tripped the boiler fan and fuel pump to reduce the load on the diesel, we started the condenser pumps; and the third immediately and without ceremony wound open the throttle on the turbo to run it up to speed, which it did with a noise like a jet engine starting up. The lecky then tripped the diesel circuit breaker, briefly plunging us into darkness again, before throwing in another to connect the turbo instead (cue for us at the control panel to restart the condenser pumps for the second time). Of course the steam pressure was now on its way back down and the water level too, so we were in a race to get the boiler lit up again before it dropped too far and we were blacked out once again.

Firstly, the turbo feed pump (steam-driven) had to be run up to restore the water in the boiler to a safe level, which took an agonising minute or so, during which time the steam pressure declined even faster. As soon as the boiler man could see a level in the gauge glasses he could then restart the fuel pump and forced-draught fan and light up a burner again. On this first attempt, however, we were just a bit too slow. The forced-draught fan had a very powerful motor – 150 hp I think – and the sudden load on the turbo, which was already trying to run on reduced steam pressure, was too much for it and it tripped off the board, giving us our second blackout of the day.

I seem to remember that we didn't have any better luck the second time round, either. Then at the third attempt, by which time we were all shattered and drenched with sweat, we finally made it. What a relief it was to get the ventilation fans running once more! The second sent half of us up topsides to have a blow, while the remainder carried on with the business of firing up the second boiler and getting the ship under way again.

The conduct of our absentee chief engineer, who in common with all my previous ones, hardly ever showed his face in the engine room, left me feeling surprised and also quietly disgusted. The passage of time, however, has given me a more sympathetic perspective. They were all in their late fifties or early sixties and probably felt they had done their bit and that the youngsters could therefore take the strain. Some at least had been through the war and seen a lot of action – both the chief and the captain on *Warwickshire* had a row of medal ribbons on their No.1 uniforms on the odd occasions I actually saw them wearing them. As I said in Chapter 5, it was the chief officer and the second engineer who ran the job anyway, and with the exception of the second on *Preston,* they all struck me as being extremely competent, and might even have felt annoyed if the captain and the chief engineer had started looking over their shoulders.

As to the voyage itself, my discharge book tells me that I joined the ship in Fawley and from there we went to Amsterdam, where we had a change of articles and the book got stamped up again. I enjoyed another good run ashore in the town, this time to a club and bar on an old ship in the harbour, before we sailed for the Gulf for our next load.

After that we headed off across the Indian Ocean to Bombay (now Mumbai). The weather for virtually the entire passage from the Gulf was hot and humid with glassy seas; there was no distinct horizon and the sea seemed to merge into the sky. This of course, made it impossible to get any accurate sun or star sights – and so, to put it bluntly, we were lost. By dead reckoning we were still at least 30 miles from the coast, but as the sea started to change colour from blue to muddy green with bits of floating weed, we guessed we must be a lot closer than that. The radar didn't seem to be much good at picking up the flat coastline, either, and by the time it did we found ourselves some 15 miles away from our intended landfall. These days the GPS would have given our position to within 100 yards while the sextant stayed tucked up in its box – but where's the fun in that?

Our terminal was actually nowhere near Bombay itself, and there was no shore leave. We were visited by a load of bumboats, however, and the ship was soon swarming with pedlars selling everything from carved elephants to women. I steered clear of the latter, but for my parents I bought a beautiful mahogany coffee table with a view of the Taj Mahal inlaid into the top and with each leg carved in the shape of an elephant's head and trunk, complete with mini-tusks. I also got some very cheap tropical shirts, which lasted quite well, and a set of top-quality cotton towels, which we were still using in 2009! I was worried that the table, which had to be packed up and posted home for me, might actually never arrive – but when I got home on leave, there it was in the sitting room. My parents, who'd had no idea that it was coming, were thrilled to bits.

From Bombay, we returned to the Gulf to load again, and then proceeded to Europe – can't remember where. As we passed Cape Town on our way back, some of the crew were being paid off, having completed their tour of duty. One of them was a junior engineer, and when we had a look at the radio message giving the names of the lucky ones who were going home and their not-so-lucky replacements, we could see that this junior didn't actually have a relief, which would have left us undermanned. The second was standing next to me when we scrutinised the list, and I asked how this had come about, to which he replied, 'He doesn't need a relief because I have recommended to head office that you take his place and they've agreed.' He went on to say that he had been impressed by my knowledge of the plant (largely thanks to JH) and my performance during the blackout and was therefore confident I could do a junior's job now.

After all the negative comments I had received at college and on *Yorkshire,* it came as a very pleasant surprise to find someone who thought that I might actually be any good, so I was delighted at this news and sent my parents a telegram to let them know. At this time I still needed another five months' seagoing experience to complete my cadetship, so I was going to finish early. Apart from anything else, this meant my salary would take a big jump from the £260 a year, I think it was by then, right up to £850, which was quite respectable in 1968. I would even be able to buy another car, having sold the MG when I first went off to sea in order to pay back the money I owed my parents.

As the most junior junior I was put on the 8 to 12 watch, almost for the first time since I had started at sea, which at least meant a decent night's sleep for a change. The reason for this was that if something started to go wrong on this watch, help was more readily available without having to get someone out of bed first. In the engine room it was generally kept by the fourth engineer, while on the bridge it would be the third mate if there was one, otherwise the most junior of the second mates.

Apart from the blackout incident and my promotion, the only other thing I can remember about my time on *Edinburgh* was when I paid off myself, at Mina Al Ahmadi in Saudi Arabia on our return to the Gulf. The customs post was simply a wooden shack at the end of a long jetty, down which we had to struggle with our luggage and then stand around outside in the blazing sun while the customs officials spent an interminable time scrutinising our discharge books and stamping them up. It is interesting to note that back in those days, seamen joining and leaving ships could travel quite happily without a passport or visa, and often with less hassle, as long as they had their discharge books. I had actually got myself a passport before I joined *Yorkshire*, but the captain, who was an old hand at travel in these parts, advised me to keep it out of sight and just present the discharge book for scrutiny unless things became really difficult. In the event the discharge book was just fine.

The trip in the taxi to the airport was interesting to say the least, as the driver didn't seem to know which side of the road he should be on. The road was just a strip of black tarmac through the desert and was mostly dead straight, so vehicles coming the other way (and fortunately for us there weren't very many) could be seen from a long way off – it was terrifying to watch them approaching without knowing on which side our driver was going to pass them. To emphasise the danger, there were numerous crashed and burnt-out wrecks beside the road as silent witnesses to previous mistakes.

I have always been interested in aviation and was looking forward to my first flight. The aircraft was a Boeing 707, and I shall never get forget that first great thrust in my back when the pilot opened up the throttles and we took off. It was a beautiful sunny day (is it ever anything else in the Gulf?) and as we climbed we could see everything laid out below like a giant map. Eventually, we reached our cruising altitude of 32,000 feet and I was very surprised to find that even from 6 miles up it was possible to pick out individual cars on the roads.

We had one intermediate stop at Rome before going on to Heathrow, where we arrived in the late afternoon. I had a great deal of luggage, including my much-travelled banjo, and didn't fancy struggling back by tube to Victoria and thence to East Croydon on the train, so I got a taxi the whole way home. As I stated earlier, Esso never queried any hotel and travel expenses, and also paid the excess baggage charges that were usually incurred whenever we had to fly to or from its ships – I seem to remember that the third engineer and his wife, who were also on the flight home with me, had over 100 kgs between them. I had not told my parents that I would be coming back home from the Gulf, in case my relief didn't turn up and I had to stay on for a bit, so they had a wonderful surprise when I walked in on them that evening.

With my newly acquired wealth, I didn't waste much time before getting another car, and after scouring through *Exchange & Mart*, I located another MG TC for sale in Wimbledon, which was not too far away. My dad drove me over to view it and after a bit of haggling it was mine for £165. This one had the original bench seat in the front, which meant you tended to slide around a bit when cornering, and I was never quite so in love with it as I had been with the first one. Still, it was a set of wheels and got me out and about visiting my friends again.

One of the strange things about returning home after being out of circulation for four or five months was the way some of my friends greeted me. After the initial 'Hi John, how are you doing?' they would then almost immediately go on to enquire 'When are you off again?' I am sure they didn't realise that this might have sounded somewhat unfriendly; indeed, these same people are still my closest friends some 40 years later – it just seemed to be the way their minds worked.

I had a great time revisiting all my old haunts and tinkering with the new car, but all too soon my two months' leave was up and I was off to sea again. I didn't know it then, but my next trip would also be my last with Esso Tankers.

9 HARD tIMES ON DURHAM

Esso *Durham* had started life in 1959 as a 36,000 dwt tanker with the bridge amidships, but in 1961 she suffered a massive explosion which blew a hole in her side. The damage was so severe that the decision was taken to cut her in half and put in a new 40-foot long section, which increased her deadweight to 40,929. At the same time, her bridge superstructure was removed and a new all-aft accommodation block put on instead. It was in this form that I joined her in April 1969.

If I had thought that the automation problems on *Edinburgh* were bad, *Durham* was even worse, and it really needed three officers on watch down below to keep up with all the running round that was required. When on stand-by a further three were required – the second engineer (who we nicknamed 'Batman') at the control platform, the electrical officer at the switchboard and the extra third in the boiler room. The second seemed like a reasonable chap, although I was to find that he was rather too full of his own self-importance and a bit of a panic merchant.

A hint of this behaviour came quite early on in the voyage. His usual practice when on stand-by was to pace up and down in front of the control console with his hands behind his back, looking supremely confident and in total command of the situation. About every third pass down the gauge board he had a habit of giving the main condenser vacuum gauge a light tap with his knuckles before resuming his pacing. As we always strived for every inch of vacuum we could get, this gauge would normally be reading 27–28" of mercury, right up near the end of the scale (the maximum possible is 30"), so it was prone to sticking – hence the tap as he went past. On this occasion we had been on stand-by for nearly half an hour, and the gauge had already received quite a few taps when nothing had happened, so when Batman gave it yet another and it dropped suddenly to 18" it didn't register with him at first, and he took another couple of paces before performing the most amazing double-take and starting to flap around like the proverbial headless chicken. As it turned out, it was merely that the gland steam controller had stuck, and it took no more than a minute or so to change over to hand control and restore things to normal. Batman's self-composure took rather longer to recover.

Unlike my other ships, *Durham* was on a regular run from Rotterdam to Nigeria and we completed three round trips during the time I was aboard her – if we hadn't kept breaking down it would have been four. Right from the start it was obvious that we were going to have a rough trip because I am fairly sure we blacked out as we were leaving Rotterdam and had to drop the pick for an hour or so while we sorted ourselves out. I can't remember what the trouble was this time – feed pump tripped possibly – but I certainly recall hearing the engineers' alarm for the first time. This was a very loud klaxon hooter situated in the passageway outside my cabin, whose purpose was summon all the off-watch engineers to go below when there was a serious problem. I know that I was in my bunk when it sounded off and I nearly jumped out of my skin. I think I was into my boiler suit and back down to the engine room within 30 seconds.

This brings me neatly onto my next topic, which concerns getting people up for their watch. This was known as 'putting them on a shake' and could be pretty difficult at times, especially if the man concerned had been drinking in the bar for several hours beforehand, so it's a pity there wasn't a mini version of the engineers' alarm that we could carry round with us at night and stick beside their ear for a few seconds to speed up the process. If you have just completed four hours in a noisy sweat box with the temperature nudging 110°F on the plates, what you crave most is to get out of the wretched place for a shower and a couple of cold beers. So if you have to go up and down the ladders two or three times to call a man before he eventually staggers down some 20 or more minutes late, it is ruddy infuriating. Over the years I have had some terrible struggles at midnight or 4 am trying to get men out of bed.

Strictly speaking, you were supposed to report for your watch some five or ten minutes before it officially started anyway, so you could check the log and find out what was going on. I can remember one occasion when I was a few minutes late myself and I queried something in the log, to which the off-going engineer replied, 'If you want to criticise anything that's happened on my watch, you tell me *during* my watch – it's nearly ten past the hour now, so it's your problem!' He was quite right, of course.

One of the things that plagued us on *Durham* was a shortage of distilled water for the boilers. In theory, seeing as all the steam used fetches up back in the condenser where it gets turned into water again, the daily consumption should be nearly zero. In practice, however, things were different: the oil burners in the boilers used steam to atomise the oil spray to improve the combustion, and this went straight up the funnel, thus possibly getting through as much as 4 tons a day. Add to this the steam used during soot-blowing (more about this in a later chapter), which was also lost up the flue, plus a multiplicity of small steam and water leaks from various joints and glands. All this meant that the normal daily consumption of distilled water on a turbine ship of this power would be between 10 and 12 tons per day. On *Durham* though, we seemed to suffer from so many disasters such as condenser leaks and boiler tube failures that our daily consumption was usually around 20 tons, and we were always playing catch-up with the water supply. Although we had freshwater tanks which we could top up in port from the mains supply, we didn't trust the quality of the water in Nigeria, so our distillation plant also had to be able to produce some for our domestic use at times. We tried to avoid having to do this if at all possible, because distilled water is awful to wash in – it feels like thin oil and however much you rinse off the soap, you never seem to be able to get rid of it.

We had two evaporators on *Durham*, each supposedly able to produce 15 tons per day, but if they ever managed 10 we thought we were doing well. They were in continuous use and required almost continuous attention. As I explained earlier, when you boil ordinary water it causes scale to be formed, which will build up on the heating surfaces and reduce the efficiency. If you are boiling sea water to start with, this problem is very much worse. So in an attempt to get round it, evaporators make use of the principle that water boils at much lower temperatures when under vacuum conditions, and at these lower temperatures scale is much less likely to form. That was the theory, but in practice we were forever having to pull out the heating coils for de-scaling, which was an unpleasant job taking several hours to complete.

Having got an evaporator started, the salinity of the water it was producing was carefully monitored to ensure it was up to the standard required for the boilers, using an electric salinometer of the kind used for checking the main condenser. This was fitted with an alarm that sounded at the control platform if the salinity got too high, and at the same time it automatically opened a dump valve to send the output to the bilge, to avoid contaminating the distilled water tanks. It was a rare occasion when we went an entire watch without one or other of these 'vap' alarms sounding off. Each time it did, someone would have to spend 20 minutes or so fiddling about with the various adjustments and settings to get the evaporator back on line again.

After the initial blackout, we had managed to get ourselves to Bonny in Nigeria without too much else in the way of drama. Bonny is a town situated at the mouth of the Bonny river, which in turn is just one of the many branches in the delta of the mighty River Niger. Bonny grew up with the country's oil industry until its importance eclipsed that of Port Harcourt, which is the older town and harbour lying a bit further upstream.

We had to anchor off for a few days while we waited for a berth. Just so we didn't get bored, we were greeted by quite a few of the local inhabitants, who paddled out to see us in a flotilla of outrigger canoes. These canoes were very slender, about 20 feet long, and appeared to have been made by hollowing out a single large log of some very dark wood. The locals brought great stalks of bananas and various other fruits that they traded for bars of soap, cans of Coca Cola (they didn't want beer) and anything else we cared to lower down to them. Even a few pairs of discarded engine room boots left by some of the previous crew were greeted with much delight. At first the locals were all happy and smiling, but after a few hours, by which time we had enough bananas to last us back to Rotterdam and our supplies of spare soap bars and Cokes had run out, they became rather surly and abusive. The only surprise to me was that they found anything at all to smile at, as they were obviously desperately poor and the climate, which I don't suppose varied much over the course of the year, was horribly hot and humid.

We were none too happy either, as *Durham* was an unbearably hot ship – and not just in the engine room, as related in the prologue. Older vessels which had been built in the days before air conditioning were equipped with fans in the cabins and a whole forest of ventilators up on deck, each with a, cowl that could be turned to face into wind to make the most of whatever cooling breezes there were. *Durham* was, however, of a later generation, which relied on the air conditioning to keep the accommodation cool and had none of these things. This was fine of course if the AC was working – but when it wasn't (and this was most of the time) then you were much worse off than if you had never had it in the first place. Our trouble was that there were so many other things going wrong that we had our work cut out just to keep the ship running and the AC was well down on the priority list. It was a Freon gas system, and due to the lack of maintenance the gas was constantly leaking away until eventually we had used up all our spare gas cylinders, after which there was no more AC.

My cabin was on the starboard side, one deck below the boat deck, and the bulkhead next to my bunk adjoined the boiler room – at times it was almost too hot to touch. This was nice and cosy in Rotterdam in winter but out here in West Africa it was hell. Nor was it possible to cool off in the shower – the domestic water pipes had to travel all the way up through the engine and boiler rooms to get to the accommodation, and by the time it reached my cabin even the cold water was too hot to stand under, when first tried. To get the temperature down enough to be bearable, it was necessary to run the cold tap for about five minutes first; with everyone else having to do the same, this was another reason why we were always short of water.

I was not the only one suffering, and we each had our own ways of dealing with it – when off watch, cold beer in liberal quantities was one answer, and the ship's consumption of this particular commodity must have been astronomical at that time. Some of the crew who had cabins on the lower decks with a porthole on the ship's side, had rigged up air scoops made from five-gallon oil drums suitably cut down with tin snips, so that they fitted neatly into the porthole opening and faced forward to catch our wind of passage. I couldn't do this because my porthole faced onto the deck and anything sticking out would present a hazard to people walking past in the dark. In the end, I found the only way to sleep was to take my mattress up onto the boat deck and lay it down in a quiet corner. Bonny was in a malarial area and we had been warned against doing this, but I was not the only one who preferred taking a chance with the mozzies to sweating their guts out in their cabin. We were provided with Paludrine tablets which we had been taking every day since we joined, and I think that most of us believed this would give us at least a sporting chance of avoiding the dreaded malaria.

Leaving Bonny, we hadn't gone very far before we had another blackout, and this time it took several hours before we could get restarted, due to the fact that without the engine room fans nobody could stay in the engine room for more than about 15 minutes at a stretch and there never seemed to be enough men available at any one time to do all the required starting and stopping of pumps and gennies etc. As with *Edinburgh*, the diesel emergency generator was far too small and could not be synchronised with the turbos. Eventually, after a particularly heroic effort by about five of us, we managed it, by which time we were all half-dead. Having finally escaped from the hellhole for a breather, we collapsed on deck and used a bucket with a long rope to bale sea water over our heads.

Although I can't remember exactly what exactly happened where and when during my four months on *Durham*, what does stick in the mind is that nearly all our troubles occurred when we were out in the tropics rather than in more comfortable climes – this I suppose, being yet another fine example of sod's law.

Typical of this was when we had a couple of boiler tube failures following straight on from the last blackout – as if it hadn't been hot enough anyway, we were now going to have to get inside the ruddy boilers. The failures were probably caused by our chronic water problems, which had meant that we had been unable to keep the quality of the boiler water up to the usual standards. Unlike a Scotch boiler, where a tube failure will be immediately obvious and potentially very dangerous, in a water-tube boiler, where the furnace volume is huge by comparison, steam from any leaking tube even at pressures of 860 psi can go straight through and up the funnel without giving any visible sign. So this time, the way we knew we had a leak was when the fourth engineer on the 8 to 12 reported that the water consumption on his watch had rocketed up from the usual 2 or 3 tons to around 12. By testing water samples from both boilers and seeing which one had lost the most treatment chemicals since the previous test, we were quickly able to establish which was the affected boiler and shut it down – thereby reducing our speed from around 16 knots to a measly 11, which was not much better than the dear old *Esso Preston*.

Our water-tube boilers basically consisted of a water drum at the bottom and a steam drum at the top (which is actually half-full of water too), each about 20 feet long, joined together by an enormous number of tubes, which also formed the roof and sides of the furnace. The water drum was only about 2 feet in diameter internally, which meant it was an extremely tight squeeze to get in, and once inside it was impossible to turn round, so if you went in head first, you had to come out feet first. The steam drum was about a foot bigger but was full of steam separators along the bottom which had to be removed and passed back out one by one as you worked your way along into the drum. These cyclone separators were a Babcock & Wilcox patent and were designed to help prevent water being carried over into the steam outlet pipe. When steaming hard the mouth of each steam-generating tube would issue a huge stream of steam bubbles into the drum, and the froth formed might eventually build up enough to cause a carry-over of water into the steam pipe, in the same way that a bottle of fizzy drink may overflow if it has been shaken up before you open it.

It took about six hours before the furnace side had cooled down enough for us to climb in and have a look, but it was quite a bit longer before we could get inside the steam and water drums to find the affected tube and plug it. Access to the furnace was provided by removing one of the burner assemblies complete with the air register, which gave an opening around 2 feet in diameter for us to crawl through. Once inside I was aware immediately of the tremendous heat that was still radiating from the metal surfaces, but there was also a nice bit of natural draught coming through which made it bearable.

The furnace was big enough to walk around in and in fact you needed a step-ladder to be able to inspect the roof. The floor was of refractory material and the sides and roof were formed of very closely spaced tubes – so close that they are called water walls. On the inboard side the tubes were spaced more widely, to enable the furnace gases to pass through before they reached the superheater. After this they continued on up through the economiser and eventually into the uptakes. These more widely spaced tubes were called screen tubes because they protected the superheater from the radiant heat of the burners. They, being exposed to the hot gases on all sides, were the most likely source of our problem. I looked around very carefully but was unable to see anything that looked like it might be a leak but Batman, with his more practised eye, spotted a slightly less soot-blackened area up by the steam drum, which he guessed would be the spot.

I think it was about this time that I first made the acquaintance of Board of Trade Lime Juice. It is fairly well known that during the 18th century the British Navy discovered the link between a lack of fresh fruit (especially citrus fruit) and scurvy, which had until then killed thousands of sailors on long voyages. Thereafter the Navy supplied all its ships with lime juice, which the crew had to drink or be flogged. It was when the Americans, during their war of independence, heard about this that they started to call the British 'limeys'. What is less well known, however, is that lime juice was still required to be provided in ships' stores well into the 20th century. No doubt all my earlier ships had carried some as well, but it was only on *Durham* that I ever saw it actually being drunk. It was brown rather than green as you might have expected, and used to come in Winchester quart bottles. It was so concentrated that I am sure it would completely dissolve an old penny, never mind take the tarnish off. However, our cook, who was very sympathetic to the torments we were enduring, had come up with a way of making a refreshing drink out of it. He would pour about a cupful of the stuff into a 2-gallon bucket of water and add a pound or so of sugar. After a good stir this mixture was left in one of the chill rooms for an hour or two. When it arrived down below, five or six of us could drink the lot in a few minutes. When the cook heard of the success of his concoction, his buckets of lime juice were provided for us every couple of hours during a work-up, and were most welcome.

When the boiler had been drained down and the pressure was off, we could remove the manhole covers at the end of the steam and water drums, which speeded up the cooling-down process. These manholes were oval and measured a meagre 15" by 10", which would mean that at least half today's adult population in the UK would probably be unable to get through them. This does indeed sound impossibly small but by stretching one's arms out straight in front of your head the shoulders go through quite easily. Hips are a different matter, however, and anyone measuring much over 36" in this area would probably get stuck. In those days I only weighed in at just over ten stone, so I had no problems – which was probably why Batman suggested that I should have first go in the water drum.

Now that we had an idea of the approximate location of the leak I was given about 20 wooden bungs and told which tubes to try first. They were all near the far end of the drum at the top, so I had to wriggle my way down quite a way to reach them and then turn on my back so I could gently tap the bungs into the tube ends above me. The next stage was for another man to crawl along into the steam drum, taking with him a freshwater hose with a tap on the end to fill the tubes I had plugged – needless to say if he got any down the wrong tube it would run straight through and pour out all over me. He would then proceed to fill all the plugged tubes – obviously the one that did not stay filled was the one with the leak. Our first attempt was a failure as they all held water, so now I had to remove the bungs and shift the search area. You can probably guess what happened next: being unable to get out of the way, every time I knocked out a bung I got a free shower with a gallon or so of none too cool water.

Batman, who was actually not a bad bloke despite his tendency to panic, had been keeping an eye on me from the outside. He now enquired if I was okay to have another go, and upon receiving my somewhat reluctant affirmative, he suggested which lot to try next. Fortunately he had guessed right this time. The offending tube would be plugged with fine tapered steel plugs well hammered in – the boiler pressure itself would also be holding them in place, so there was really no possibility of their ever coming out on their own. Batman insisted on doing this part himself and I was more than happy to let him do so. I wriggled backwards down the drum again until I was able to stick my feet through the hole, and he grabbed hold of my legs and gave me a good heave to help me out. Having been released from my cylindrical and claustrophobic prison, soaked to the skin and with a face as red as a beetroot, I was sent up topsides for a much-needed blow and a dry boiler suit.

Returning to the engine room some 20 minutes later, I was surprised to find Batman running round in a terrible stew, exhorting everyone to hurry up and finish the job so we could get the boiler going again. It transpired that the other boiler had also now sprung a leak and we had to get the newly repaired one on line as soon as possible, before we lost too much water. The poor chap had been going a bit thin on top before he joined *Durham*, but by the time we paid off he needed a wig.

As soon as we had the manholes back on again, we started filling the boiler, and the second a water level appeared in the gauge glass, the burning torch was stuck through the lighting-up hole and the oil and steam atomising was turned on. When the burner ignited with its usual whoompf, the air register was opened and the forced-draught fan adjusted until we could see through the periscope that the smoke had cleared. We raised steam in that boiler from nothing to 860 psi in about 25 minutes. So now, having just completed one major work-up, we were straight into the next one without a break. By the end of that we were all just about finished, both mentally and physically – so much for the joys of steamships.

The quality of one's fellow colleagues when working in such adverse conditions as these make all the difference, and I had one very good mate on *Durham*. Peter Brinkley was his name – a big man with a big heart. Whenever we were off duty at the same time we would share a few drinks and compare our respective tales of misery, which we seemed to find strangely cheering. I can also clearly remember several occasions when we had come up in the early hours of the morning and would stand by the rails on the poop deck in our tattered and oil-stained boiler suits, drinking cans of beer and watching the sun come up. Being quite a bit heavier than me, he was very often one of the steam drum men when doing boiler jobs, while I used to do most of the water drum work, and we often used to chaff each other about who had the worst job.

Just to make a change from stories of working in the heat, the next incident I shall describe involved cold water in abundance. *Durham* had been a state-of-the-art oil tanker when built in 1959, but for some reason she still had an old-fashioned tail shaft (see glossary) of about 20" diameter running in lignum vitae bearings, with a simple stuffing box and soft packing to keep it watertight. Lignum vitae is a very hard and dense wood (so dense that it does not even float) and which requires no lubrication other than sea water when used as a bearing, whereas modern ships have the shaft running in white metal bearings with oil lubrication and rubber lip seals to keep the water out.

On *Durham* the stuffing box had leaked from the day I joined, and it gradually got worse. Every so often we would pull up the gland nuts to tighten the packing and reduce the leak but the shaft itself was worn (lignum vitae will wear down the steel shaft rather than the other way round, due to particles of grit etc. becoming embedded in the wood, which converts it pretty well into a fine file) and it would usually be leaking again as badly as ever within a few hours. Eventually we had pulled up the packing as far as it would go, after which the leak escalated rapidly, so that soon the sea water was coming in fast enough to fill a bucket within seconds. This meant we were continually having to pump the shaft compartment bilges out – sometimes for an hour or two at a time. Eventually, our overworked reciprocating bilge pump threw its hand in, so now we really were in trouble.

We had other pumps we could use – the general service pump for one – but this was of the centrifugal variety, which are notoriously difficult to get started unless there is a positive head of water on the suction side. This one was pretty clapped out anyway, and however much we primed and fiddled, it would never keep pumping for long. Meanwhile the water in the shaft compartment was steadily getting deeper. Our electric oil fuel pumps for the boilers were also in this compartment and it was imperative that we kept them dry or we would be finished, so we were running round frantically wrapping them up in rubber sheeting to keep the water away from the motors. Finally we were forced to let the water overflow through the watertight door into the engine room. This could only be a temporary measure but it gave us some more time to plan what to do and stopped the level from rising still further and swamping the oil pumps.

Eventually we decided that we would have to stop at sea and attempt to repack the stuffing box around the shaft. As this little drama was unfolding we were heading south in a westerly gale near the bottom end of the Bay of Biscay, and the bridge informed us that we had about three hours to do the job or we would be driven ashore. This, we thought, would be just enough time, and we stopped the engine and hove to. The deck department ballasted the ship well down by the head in an attempt to get the propeller shaft above sea level and reduce the leakage – but as soon as we were stopped the ship drifted around until she was beam-on to the waves, which made her roll quite heavily. This caused all the bilge water in the shaft compartment, which was by now some 6 feet deep, to thunder about from side to side with the roll of the ship, so that at times the tank top on the uphill side would be dry while the other side would be under 10 feet of water.

The chequer plate walkways were raised above the tank top by about 6 feet and were held down onto their angle iron bearers by countersunk brass screws – about 3/8th Whitworth thread, I would guess. As the water crashed up against these plates from the underside, these screws started to shear off – rather like a cartoon character who has eaten so much that his shirt buttons start pinging off. Quite a few screws were missing anyway, and eventually a couple of the plates themselves worked free. As I was wading down the walkway with a handful of pump spares I failed to notice that there was a gap in the plates in front of me and promptly fell through it into the freezing and dirty water. Batman was right behind me when this happened and when he saw my head disappear from view he stuck his arm down, grabbed me by the neck of my boiler suit and heaved me out again. Amazingly, I still had hold of the spare parts in my hand, which relieved him greatly. I have to admit though, that I was badly frightened by now and had no idea whether or not we were going to get out of this particular scrape.

The first thing we did was to replace two of the stuffing box gland studs with longer ones, so that whatever else happened we would always have two nuts in place, to be able to pull the gland back into place if the water pressure proved to be too high. This done, we took off all the other nuts and started steadily unscrewing the two on the long studs. The water pressure obligingly pushed the gland up along them until it was clear of the stuffing box. A bit further still and the remains of the old gland packing were also pushed out, so we didn't have to waste time fishing around with packing extractors to remove it. Now that there was no packing at all, every time a wave passed under the stern of the ship, a solid jet of cold water shot in all round the shaft.

The next stage of the game was to fit the new lengths of packing, which we had previously cut to length, into the stuffing box. This packing resembled a very greasy square rope. It took four of us to repack that wretched gland, two of whom (including me, as I was already soaking wet) had to stand in the bilge to either side of the shaft while we were doing it. We were up to our chests in the water and within a few minutes were shaking with cold. At first, every time we managed to get a length of packing in, the next wave of water pressure would promptly spit it back out again, and we were getting nowhere. Somebody then suggested we try shoving it in with broom handles, which would get it further down the stuffing box and give us a bit more purchase. This idea seemed to work and after waiting for a suitable gap between the spurts of water, we finally managed to get the first piece in. Then it was a case of holding on like grim death to keep it there until the next wave had passed and we could insert another – and so on.

I can't remember how many pieces had to be put in, but we had to be sure that the stuffing box was completely filled if we were not to have wasted our time. The trouble was, the more we put in, the harder it got to shove the packing down the hole. Eventually the third, who was standing on the opposite side of the shaft to me, reckoned he could feel that it had reached the end of the stuffing box, so we could start nipping up the two nuts to bring the gland back down the long studs. The last few inches were particularly tricky as we could no longer bring our broom handles to bear on the packing and had to use them on the gland itself, to quickly take up this final gap before any packing got shoved out again. As soon as we had all the nuts on we were safe, and immediately phoned the bridge to get them to start pumping out the ballast so we could get going again – they reckoned another half-hour and we would have been on the rocks!

We didn't tighten up the gland immediately to completely stop the leak, in case the friction overheated it and undid all our good work, but after we had been running for about half an hour to give it time to settle we very gently started nipping it up until the leak was down to about a pint a minute. We were more than happy with that, as this would be enough to keep the packing lubricated and cooled, while the quantity coming in was negligible compared to what it had been. By this time also, the fourth and his junior had repaired the bilge pump, and we were finally able to get rid of the water and heave a sigh of relief. There was quite a party in the bar that night to celebrate our escape, I can tell you!

And so the voyage continued: we seemed to lurch from one crisis to another in a relentless procession, and if we ever managed a whole week of simple watch keeping without a panic we considered it to be good going. I can certainly remember at least two more boiler-tube jobs and several blackouts for one reason or another – feed pump failures were favourite I think but eventually we arrived in Bonny for the final time. We were all looking forward to loading up and getting back to Rotterdam, where we would be paid off. Esso had sold the ship to a Greek company and so this would be her last ever voyage under the Red Duster- not that any of us could have cared less, as all we wanted was to get off the ruddy thing.

We had to anchor offshore waiting for a berth as usual, but eventually the bridge told us that we would be going in and could they have steam on deck to operate the windlass and winches. On *Durham* the exhaust steam pipe returning from the aft winches dropped down through the decks into the shaft compartment before making a long horizontal run back to the auxiliary condenser in the engine room. There was a valve down there which had to be opened first, but on this occasion it had been overlooked. We soon got a phone call from above to say that although there appeared to be steam at the winches they still wouldn't run. Batman immediately twigged that we had forgotten to open this valve and sent me down there to do it. It was situated a few feet from the deckhead and to reach it I had to climb about five steps up a ladder to a small platform. I climbed up and commenced to open the valve, but the pipe immediately started jumping and banging with violent water hammer. This is a phenomenon that occurs when a steam line is badly drained and pockets of water are lying about. As soon as I started to open the valve, which had steam on one side and vacuum on the other, the steam was picking up these slugs of water and hurling them down the pipe, so they were impacting hard against the first obstruction they met, which happened to be this particular valve.

I didn't like the sound of this one bit, so I rapidly shut it again and returned to the engine room, where I asked Batman if we could shut the steam off first. He retorted that we hadn't got the time to spare for that and to just go down again and get on with it. Doing as I was told, I returned to the valve and started slowly opening it up once again. There was more horrible crashing and banging and the whole pipe was jumping around on the support hangers. Suddenly there was an extra-large bang and everything seemed to just explode in my face. I was blown backwards off the platform and down onto the plates, where I found myself lying in a pool of hot water with the complete valve cover and spindle assembly in my hand. Mercifully, the water was not quite boiling and because I had been standing right behind the valve cover, my face had been shielded, but my left arm from the shoulder down had taken a hefty dollop and was now hurting like hell. I returned to the control platform where Batman was jumping up and down in a panic, wondering where the vacuum had suddenly gone. I told him rather less than politely what he could do with his vacuum and threw down the remains of the fractured valve at his feet. It was only then that he realised I had been hurt and sent me off up to get some first aid.

Back in my cabin I peeled off my boiler suit to find that virtually my whole arm from the shoulder to the wrist was now an angry red with large blisters starting to form. The best thing for burns and scalds is cold water, but of course I had to run the shower for the usual five minutes before I got it, which didn't help. After standing under the shower for about half an hour, which eased the pain considerably, I went and found the chief steward, who was the ship's first aid man. He bandaged up the affected areas, gave me some painkillers and told me to come back again in a few hours' time so he could see how the blisters were doing. After this traumatic experience you might think that I would have been entitled to have a few days off. Did I get them? Like hell I did! Batman had me back down on watch again that evening. As a small consolation I was put on light duties, which basically meant that all I had to do was mind the shop and get the log in.

On reporting to the chief steward again a bit later, he unwrapped my arm to find that I now had two enormous blisters together with several smaller ones spread over the biceps and forearm. He made a small incision into each of these with a sterilised knife and drained off about a pint of fluid, before applying a fresh dressing – a procedure that had to be repeated twice a day for several more days. He went on to tell me that although the skin over these blisters was already dead, it would continue to serve a useful purpose by keeping out the dirt and he warned me not to try removing any of it. This proved to be good advice, and within a couple of weeks I was able to dispense with the dressing altogether and the arm was starting to look pretty healthy again, albeit with large flakes of dead skin peeling off and the new skin underneath looking very pink and tender. Today, there is no sign at all that I once lost nearly half the skin on that arm.

Meanwhile, until we could make up a distance piece to take the place of the valve that had fractured, we were unable to supply steam to the aft deck and had to go back to the anchorage. This meant we lost our slot and it was nearly a week later before we finally got a berth.

When we eventually tied up and got the ship's mail, which was always eagerly awaited, it was only to bring me more bad news. My cat, Spud, had become ill and my parents had had to have him put to sleep. As I sat in my cabin and read that letter I don't think I'd ever felt so low in my life. I was overworked, exhausted and still in pain from my arm – and now my dear old cat was dead, too. I wasn't sure what else I would do in the future but I did know for certain that it wouldn't involve another trip with Esso Tankers. My mate, Peter, had also had enough and told me he too intended to resign when we got back.

Nothing else terribly dramatic happened on our final trip back to Rotterdam, but what I do remember is the deterioration in our usually excellent food. There was a rumour going round that the chief steward had done a deal with the new Greek owners and sold off some of the ship's stores in advance. I can't say I really believed this to be true and think it more likely that someone in the office had been wielding a red pen over the store's orders. There was no doubt, however, that the variety of dishes on the menu and the size of the portions had both taken a turn for the worse. This meant that the fridges and store rooms got rather more than their fair share of attention when we were doing the rounds each night, and somehow a tin of peaches or a fresh mango always seemed to fall off the shelf as we went past and had to be rescued.

When we got to Rotterdam, everyone apart from a few engineers, one second mate and the lecky were signed off articles and went home straight away. The reason that the remaining few (and I was one of them) stayed behind, was that someone from Esso had come down to meet the ship and promised us large sums of money to keep ship until the Greek crew arrived and then spend a few days showing them how everything worked. The third engineer and the lecky even volunteered to do a full trip with the Greeks, having been offered a salary three times more than Esso was paying them. They asked me, too, but I'd had more than enough of *Esso Durham* and just wanted to go home.

This final week in Rotterdam passed quickly enough, and what I remember most about it is that the new Greek cook somehow managed to start serving up some superb meals again with whatever stores had been left behind for him. Eventually the rearguard too was paid off and we took a taxi to the airport to fly home. So that return to England, in August 1969, was the end of my time with Esso Tankers – and possibly even the end of my career at sea, as far as I could tell at that time.

10 UNIVERSITY, FACTORY WORK AND SOME DREDGING

After a week or two of general relaxation and trying to forget about the miseries I had endured on *Durham*, I had to start thinking about what to do next. I didn't want to change course completely, as that would mean my whole apprenticeship would have been wasted. At first I started looking around for a position with a general cargo company, as I fancied having longer periods in port and seeing somewhere more interesting than the Gulf. This proved fruitless, however, as the general cargo trade was already in decline: ships were being sold and their crews laid off. Those that remained offered virtually no chance of promotion, and in some companies it was not uncommon to find fourth engineers who had chief's tickets and second mates who had their master's.

Some time later I was chatting in the pub to one of my former schoolmates who had gone to university, and I mentioned that perhaps I should have tried a bit harder at school and done the same. He replied that since I had an OND in engineering already, it should still be possible to do it, and that I could be accepted as a mature student. I found that Surrey University ran a mechanical and marine engineering course, and that they would be only too happy to see me for an interview. I duly presented myself to the head of the engineering department and after quite a brief interview, during which we chatted mostly about some of the things I had done at sea, he offered me a place on the spot. He did warn me, however, that my lack of A-level maths would probably make things quite hard for me for the first few terms – how right he was!

When this interview took place, it was already halfway through August, and the deadline for getting in the UCCA (now UCAS) admissions forms long past, but he gave me the forms anyway, told me to return them directly to himself and he would sort it out, which is what happened. I had saved a bit of money from my salary and also got a couple of grants, the main one being a fairly substantial mature student grant supplied by my local authority (how things have changed). I was also awarded £100 a year by the Institute of Marine Engineers, of which I was a student member, which meant I would be okay financially and would not have to lean on my parents.

My dad was already 72 and only worked part-time now, as did my mother, who was just a few years younger, so they were both pleased that I would be able to see more of them and also relieved that I wasn't going to have start borrowing money again. I think they were quite proud too, that a Richardson was finally going on to university. Both my parents were extremely clever, but circumstances had denied them their chance of further education and they had ended up doing quite low-paid jobs – my father as an insurance agent then a book-keeper, and my mother as a nurse in a mental hospital.

The main campus at Surrey University is in Guildford, right next to the cathedral. It was just a bit too far for me to commute every day, and because of my late admission there were no places available in halls, so I had to get some digs. The university had an office dealing with lodgings for students not in halls, and the first place they sent me to was a newly built private house on a smart estate a couple of miles away, occupied by a married couple with two young children. I stayed there for a few weeks but I don't think they appreciated the way I was in and out at odd hours, nor my scruffy old car sharing their pristine drive. Once or twice I popped back for lunch to find the lady of the house still floating round in her skimpy nightdress, which I found somewhat disconcerting; another time when I walked in I disturbed the pair of them in the middle of a furious row. After this the atmosphere was never terribly relaxed, and they suggested that I might be better off if I found a place a bit nearer the campus.

I contacted the accommodation office again, which gave me another address to try. This turned out to be a great big Victorian house in an expensive suburb, occupied by the Archdeacon of Surrey, his wife and two daughters. The archdeacon was a jovial type with a booming voice and a very upper-class accent. His wife was terribly quiet and hardly ever spoke at all except to answer a question, while the daughters were both very attractive and extremely snooty – they obviously regarded an engineering student as some lower life form. It was either that, or else they felt it to be demeaning in some way that the family was having to take in a lodger.

At breakfast time, which was the only occasion when we were all together, the daughters used to ignore me completely and carry on a conversation as if I was invisible, the wife never spoke anyway and the archdeacon was too busy opening his post and munching his eggs and bacon to spend any time chatting, so it was hardly a sociable meal. Just once, however, the proceedings were considerably enlivened: the two daughters were going on about some party they had been invited to while the archdeacon was quietly reading his post as usual. Suddenly he threw down the letter he'd been reading and burst out: 'Damn and blast it! Some bugger's stolen the lead off such and such church roof – and it's the second ruddy time this year!' One daughter immediately piped up with 'Father, do mind your language, especially as we have a guest,' which at least proved she knew I existed. This gave me my one and only chance to join the conversation: 'Don't mind me – if he'd ever been to sea he might have learnt a few worse things to say!' I can't remember what was said next, except that the daughters seemed terribly embarrassed by their father's outburst and both got up and walked out in a huff. The archdeacon and I got on quite well after that.

As for university life – well, I can't say I enjoyed it at all. Not being resident on the campus, I was excluded from most of the evening social activity and in any event most of my fellow students were spotty-faced youths straight from school with whom I had very little in common. The work itself I found extremely difficult, and it appeared to bear no relevance to the kind of engineering skills I was more familiar with. Whether the subject was fluid mechanics, strength of materials, or working out the stress in complex shapes, it was all maths – mostly calculus, for which I was ill prepared. This relentless obsession with maths pretty well took up the whole of the course time. There was no practical experimentation work and nothing, that in my view, would have had any practical application. During my entire 29-year career in the Merchant Navy, and even afterwards when I became an engineer surveyor and had to work out stress calculations for various pressure vessel components, I never once came up against a problem that required calculus to solve it.

By the end of the first term I found myself wondering whether my life had once again taken a wrong turning but I decided to stick it out for at least the full year to see whether or not my mathematical skills would improve enough for me to get a grip on the course. In the end, however, my mind was made up for me. I was attending a class one morning a few weeks into the new term, when the head of department walked in and had a whispered conversation with the lecturer, who then spoke up and asked if Mr John Richardson was present. I duly stood up and the head of department came over and said I should pack up my work and come along to his office. He looked very grim and I had no idea what I might have done to deserve being hauled out of the class like that, so as soon as we were in the corridor outside I asked him. He apologised and said that he was very sorry to tell me that my father had died suddenly and that I was wanted at home. This came as a total surprise, and for a while my head was spinning as we proceeded to his office. When we got there he was unable to give much in the way of further details, except to say that it was my sister who had called, to say that Dad had died in hospital during the early hours of that morning. The head offered to call me a taxi to take me to the station, but I assured him that I would be able to drive and wandered sadly back to my car.

The real shock of the event did not set in until I got home and embraced my poor mother but as I drove off from the campus I was quite calm and I think that even before I got back to the digs to collect my things, I had made up my mind that this meant the end of university for me. The archdeacon seemed to be genuinely upset for me and said I should not be too hasty, but in my heart I knew what I would have to do: my mum would be unable to afford to run the house on her own and so I was going to have to get another job.

Why had Dad died so unexpectedly? This was a time of huge industrial unrest with strikes by the miners, power workers and various other groups all trying to instigate change by inflicting upset and misery on the general public, and there were regular power cuts. My dad had climbed up onto the kitchen table to reset the mains electric clock after one of these blackouts, and had slipped and fallen, breaking a couple of ribs in the process. There is not much that can be done for this injury beyond strapping up the chest, which restricts the breathing. This had caused him to get pneumonia and he had then been taken to hospital. On the Sunday evening when I had last seen him he seemed to be making a good recovery, although still in some pain, and I had not felt worried when I drove back to Guildford the following morning. He had been a veteran of World War I and as I was leaving I said that he shouldn't feel shy about asking the nurses for painkillers, as he was not in the trenches now and there were plenty to go round.

It turned out that those would be the last words I ever spoke to him. He suffered a massive stroke during the night and died without regaining consciousness. My poor mother was completely shattered, and although she tried to rebuild her life and carried on nursing for a while, she never really got over his death. They had been married for nearly 25 years and had been blissfully happy – I can't remember them ever having a row. My dad was the most gentle man I have ever met, and he never had a cross word with me either. Whenever I had done something naughty or stupid he would simply sit me down in front of him and quietly suggest that perhaps I should have done things differently. I loved him to bits.

He had been a Lewis gunner during World War I and had been wounded in 1917 at the Battle of Menin Road (part of the Passchendaele Offensive), where half of his company had been killed or wounded. He had recovered well enough to be sent back again in 1918 (I can't imagine how hard that must have been) and went on to win the Distinguished Conduct Medal – at the time this decoration was second only to the Victoria Cross and rarely given to the lower ranks. He never talked about his experiences, however, and when I found his medals tucked away in a drawer and asked him about the DCM, all he would say was that for every one who got a medal there were at least another ten who should have done and didn't.

It was only after his death, when my mum handed over some fragments of his diary for me to read, that I found the story behind the medal and his citation. After manning a forward outpost with his Lewis gun section, he had found himself behind enemy lines, and had captured two prisoners. One by one the rest of his men were killed, but despite being fired on by both sides he had somehow made it back to our own lines with the prisoners. When he finally reached his own company two days later, they had already given him up for dead.

After his funeral, both my brother and sister, who were both much older than me, offered to help with Mum's bills so that I could stay on at university – but I didn't feel I could accept anything from them, especially as I seemed to be doing so badly there anyway. In the end I don't think that my sudden departure was any great loss to the academic world. Meanwhile, the local authority lost no time in reminding me to repay the balance of my student grant, although the Institute of Marine Engineers said that I could keep their £100, which was very generous.

After a week or so of scouring the local paper for jobs, I spotted an advert for an engine fitter with Field Aircraft, a firm that operated from the old Croydon Airport. Although I had no previous experience of aircraft engines, a marine engineering apprenticeship is a very broad-based training and I decided to apply anyway. They too obviously thought I might be worth a shot, because after a short interview they decided to take me on for a month's trial, and I started work the following day.

The company had three main lines of work at that time, the main one being the overhauls of Rolls Royce Dart turbo-prop engines, which is what I was needed for. It also overhauled variable-pitch propellers, which are a lot more complicated inside than you might think – C130 Hercules props being the most common. Finally there was a small piston engine section, where a few World War II vintage Pratt & Whitney radial engines were being worked on. Most of these were R1830s from DC3s (Dakotas).

Initially I was put on the strip-down section, where the incoming engines were roughly cleaned off and then dismantled. The idea behind this was twofold: firstly to give me a good general idea of how the main parts and sub-assemblies fitted together and secondly to generate fewer possibilities for me to make a mistake and damage something than during the rebuild process. The engine, having been reduced to its component parts, would then be sent off for everything to be thoroughly cleaned and inspected. At this stage some parts would be discarded, either because they had worn beyond the limits or else because they were time-expired.

After my probationary period I was obviously considered to be worth my place because the company gave me a small pay rise and transferred me to the reduction gear assembly section, where I was to remain for the rest of my time with the firm. There was only one other fitter in this section, who was a cheerful sort of chap and never minded answering my many questions. He was ex-RAF, and had worked on Merlin engines during the war.

The Rolls Royce Dart turbo-prop engine was quite a popular choice at the time for medium-range airliners and freighters such as the Vickers Viscount, the Fokker Friendship, the Handley Page Dart Herald and the Argosy. The reduction gear assembly was situated immediately behind the propeller and was quite a complex box of tricks. Apart from its main function, which was to reduce the speed of the jet turbine shaft down to something that would drive a propeller, there were four auxiliary drives connected to it via bevel gears: these operated the fuel and the oil pumps, the propeller pitch control unit and the water–methanol unit. The latter was a device which increased power for short periods by injecting said mixture into the compressor – this was mostly used to restore take-off power in tropical conditions.

It took me about two months before my mentor considered me to be good enough to be left on my own, and I began to see why I had been employed. Several key personnel were being sent out to Africa where the company had got a contract to overhaul Dart engines and train up local staff. This work was expected to take at least six months – which left me as chief (and only) gearbox assembler. I didn't really mind, as I got another pay rise and still found the work quite interesting.

It was all so different from being on board ship – everything was clean, for a start, and we all worked in white lab coats. There was never any hurry about doing anything either – it was always a case of doing a good job, not a quick one. Every stage of assembly had to be signed off by me, then checked and countersigned by an inspector before going on to the next one. There were always about five or six gearboxes on the line at once, as it was quite common for there to be parts shortages, which would cause work on a particular gearbox to come to a halt for a while. We were never allowed to pinch bits off one engine to complete another, either, unless we first went through a complicated rigmarole of sending it back to inspection to check it for compatibility and amending all the paperwork. This was because there were many different marks of Dart engine and scores of possible modifications, so that even though a part might look right, there was no guarantee that it was the right one.

Occasionally, when things were a bit slack, I would wander over to the piston engine section for a nose round. One fascinating sight was the magneto test rig which ran the mags' at full speed for several days at a time and made the sparks jump a gap several times larger than when they were installed in the engine itself. With 14 cylinders and 2 spark plugs per cylinder, that made 28 electrodes all sparking and crackling away. These American Pratt & Whitney radial engines were superbly engineered and finished both inside and out, and more than one of the fitters working on them, most of whom were also ex-RAF, said that they were much more reliable than the more famous Merlin that powered our Hurricanes, Spitfires and Lancasters – an opinion that I am sure would cause much indignation among British aircraft enthusiasts. To be fair to the Merlin, it had been designed as a wartime engine and was not expected to have much in the way of life expectancy – quite a few only lasted one mission before being shot down.

After I had been at Field's for about six months, I started to get itchy feet and began to look around for something else. It wasn't a bad job, but I just didn't fancy spending the rest of my life working in the same factory, clocking in every morning and out every night. There was nothing much to look forward to in the way of promotion either – after about ten years I could expect to be a charge hand and after another ten I might be a foreman or an inspector – but I would still be punching the time clock every day.

Life at home was pretty difficult too: my mother had done her best to get on with her life but I never really saw her smile again, and at low times she told me more than once that she wished she had died with my father. I am sorry to say that my being with her at home appeared to do little to ease her pain, so when I was offered a salaried position which would involve quite a bit of travelling and working away from home, I didn't think it would make any real difference to her, while the extra money would definitely come in useful.

I had seen an advert by Westminster Dredging in Lloyd's List, the main source of information in the shipping world at the time. What the company wanted was a trainee dredging superintendent, a shore-based position, so although I would go on to spend considerable amounts of time on various dredgers up and down the country, none of these would have an entry in my discharge book. What the superintendent does is monitor the dredging operation by estimating how much spoil is being taken, and then analysing said spoil to see what it consists of, to ensure that the vessel is working in the right place to the right depth. A superintendent also liaises between ship and shore, to keep the client updated on progress and to make sure the vessel does not stop work for want of spare parts or stores. In the event of a breakdown, the superintendent will also organise the repairs if shoreside assistance is required. What the trainee does is act as a general gopher for the superintendent.

Westminster Dredging was the British division of Bos & Kalis (now Royal Boskalis) the Dutch parent company. The Dutch were the world leaders in the dredging industry, and Bos & Kalis had its headquarters at Sliedrecht in the western Netherlands, where there is even a museum devoted to the history of dredging – this does not sound like much of a fun day out! At my interview, which had taken place in Newcastle, at least one of the panel was Dutch, and I was soon to learn that if you really wanted to get ahead in this game, you stood a much better chance if your surname was prefixed by Van or that you could at least speak the language.

I was asked at the interview whether or not I had a reliable car, as there would be a lot of travelling, and I assured them that I did – by this time the MG had been replaced by a souped-up Mini. This car, which rejoiced in the name of the 'Marigold Mini' for its unique orange paintwork, had been prepared by the Canova brothers who were (and still are) my good friends. I was a member of a car club at the time and used to take part in the odd rally or production car trial with John Canova as my navigator. This particular Mini had a 970 cc engine with all the usual modifications such as a big valve head, hot camshaft, high compression pistons and a Weber carburettor. The brakes, however, were still quite standard and could fade alarmingly during competitions.

Although the car had never let me down before, it was more suited to tearing round over shorter distances than being flogged up and down the motorways to the office in Newcastle or to Heysham harbour in Lancashire, where my first dredging job was. I used to come home every weekend I could, to check up on my mother, so the mileage soon started to rack up, to the detriment of the engine.

For my trips to Heysham, rather than simply going up the M6, which would have been more direct, I preferred to go up the M1 until I was past Sheffield and then cross the Pennines, as there was a greater choice of routes and the Mini was great fun over the steep and twisting roads. On one occasion I had decided to try the Woodhead Pass and was blasting up onto the moors above Penistone when there was a loud bang and the car immediately filled with smoke. By dint of sticking my head out the side window, I was just able to see where I was going and managed to bring the car safely to a halt in a lay-by. It was immediately apparent that the engine was not going to get me any further and I was miles from anywhere in some very bleak and empty countryside. I thought that if I could only turn the car round, I might be able to freewheel a good bit of the way back down to civilisation, which might save me a very long walk to find a phone. The road behind me was straight for quite a way, and after checking to see there was nothing else coming up, I simply released the handbrake and let the car roll backwards. After about 50 yards I was going quite fast and gave the steering a rapid full turn to the left, which neatly reverse-flicked the car so it was now facing the right way again. This was something that I don't think I would have dared do in any other vehicle, especially as a mistake would have meant the car coming off second-best to a dry stone wall – but in the Mini it was child's play.

By dint of using the brakes as little as possible, I was able to coast most of the way back down to Penistone, where I found a small garage that could rebuild the engine – it turned out to have dropped a valve and holed a piston. This was on the Monday morning, so I continued by train to Heysham and returned to collect the car on the following Friday. As I had been rendered temporarily car-less, the superintendent had to do all his own 'gophering' that week but he was a decent sort and used to take me to and from my hotel in his car and even out for a meal once or twice. In any event, the job seemed to pretty well run itself, and our work consisted mostly of going around the harbour in a launch with an echo sounder to ensure we were achieving the required depth of water.

When I finally collected my car they showed me all the broken bits in case I was reluctant to pay the hefty bill. The cylinder head had also been damaged beyond repair, so the garage had fitted a standard head from an 850 cc Mini just to get me going again. The compression ratio was still quite high on account of the smaller combustion chamber, but the car was never really the same after that so I sold it at an auction and bought a 1,275cc Mini Cooper S instead. This was quite a bit more powerful than the old Mini and it also had the newer Hydrolastic suspension, which made the ride a lot more comfortable on long journeys.

In total I spent about a month working in Heysham harbour, where my first vessel was simply named *WD54*. She was a self-propelled hopper grab dredger and had three Priestman grab cranes mounted on the deck – one each side amidships and one on the bow. She was powered by a pair of triple-expansion up and downers; they were quite a bit smaller than the one on *Esso Preston* and not nearly so clean and well cared for, but I was happy to see them all the same.

When in position, the cranes would start loading the hopper, which would be full to the waterline with sea water after dumping the previous load. As the spoil built up in the hopper the displaced sea water would overflow onto the deck and out through the scuppers. When the hopper was full, the vessel would proceed out to sea to the spoil grounds. Along the bottom of the hull was a set of large hydraulic doors that could be opened to allow the contents of the hopper to be dumped, after which they would be closed and the ship would go back for the next load.

From Heysham, we went off to Liverpool – to the Huskisson Dock, I think it was. This area had been heavily bombed during the war and on a previous visit I was told that one of the crane drivers had brought up an unexploded bomb in the grab. Most people would have stopped work immediately and dialled 999. Not this intrepid chap, though. The crew were paid a bonus based on the number of loads they dug out, so he simply slewed the crane round, gently deposited the bomb on the quayside and then calmly carried on working.

There are several other kinds of dredgers, the bucket dredger probably being the most familiar. This consists basically of an enormous continuous steel chain with digging buckets mounted at regular intervals along it – this part is called the ladder. The angle of the ladder can be varied, altering the depth it works at. The full buckets ascends the ladder until they reach the top, where they tip over and deposit their contents into a chute leading to a hopper barge alongside. Now upside down, they would return down the underside of the ladder and back into the water for another go. The dredger herself is anchored in position by steel hawsers and can be moved around as required by winches.

The one I was on was called *Elephant*, and I only spent a few days working with her, somewhere near Chichester I believe. Like *WD54* she was steam-powered, and the neat little compound engine was obviously someone's pride and joy because it was kept beautifully clean and polished. Despite the filthy nature of the job, the accommodation on board was also maintained in a very clean condition, and no dirty boots were allowed past the changing room. The crew were a happy and self-contained bunch and, I think, worked week on and week off. There were a couple of keen fishermen on board and they used to supplement their rations with fresh fish they had caught – I certainly enjoyed a couple of very fine meals with them.

After this pleasant interlude I was sent to Rotterdam to spend a few days on the company flagship, a trailing suction dredger called *Prins der Nederland*, which at the time was the largest of its kind in the world. On these vessels, great steel pipes on flexible joints are lowered down the side of the ship until they reach the seabed. The ends of the pipe are splayed out so they can suck an area several metres wide: these are shaped rather like the carpet sweeper attachment on a vacuum cleaner. Enormous dredge pumps then suck up the mixture of water, mud, gravel etc from the bottom. The vessel steams up and down over the prescribed area trailing the suction pipes behind it, and the dredge pipes discharge into the vessel's own hopper, thereby displacing the water in the same way as described for *WD54*. In those days *Prins der Nederland* was almost exclusively employed in dredging the Nieuwe Waterweg ('new waterway' – which had actually been opened in 1872!) leading from the North Sea to Rotterdam. Rumour had it that after her first year or so in service *Prins der Nederland* had dropped so much spoil in the designated dumping area that this too had become a hazard to navigation for the largest ships.

I think this little trip on *Prins der Nederland* was really to show me that the company operated more than just old steam-powered relics and that it was leading the way with this new breed of dredger. She was the first large diesel-powered ship I had seen, and had a lot of automation for those days. I was very much superfluous to requirements on her, and it was only a week or so before I got sent back to *WD54*, which had now steamed all the way up to Aberdeen to commence dredging the docks. I was to spend the entire winter in the Granite City, but I was very comfortably installed in the old but comfortable Douglas Hotel near the centre.

Westminster had a small office right on the quayside, where I usually started my day, getting my orders from the superintendent. Occasionally he would go away for a few days and leave me in charge, not that this was terribly demanding as it consisted mostly of chasing up outstanding spare parts and making sure the crew members got all their groceries. I also had to estimate the amount of spoil we removed and agree the figure with the harbour master's office – this was really just a matter of counting the number of loads the vessel took out to sea and guessing the percentage she had been filled each time.

After we had been dredging one part of the harbour for some days – I believe it was called the Torry Dock on account of it adjoining the suburb of that name – it became increasingly apparent that the depth of water was not increasing at the rate it should have done, if my figures for the amount of spoil we were removing were to be believed. The harbour master naturally took the view that the figures must be wrong and there were some quite heavy discussions between him and my boss, who had checked them for himself and was backing me up. After one more day's dredging the reason for this apparent discrepancy revealed itself in a rather spectacular fashion. On the quayside adjacent to where we were working stood a row of large oil storage tanks and the plant operator had come down that morning to find they were all leaning over alarmingly toward the water. Some pipes had fractured at their bases but fortunately the valves had been shut and no oil had escaped. It seemed that the harbour bottom in that area consisted of fine silt, and as fast as we had been digging it out, more had run in from under the jetty until the foundations had been undermined sufficiently to cause the collapse. After that there were no more arguments with my figures.

There was a very pretty girl called Betty, who worked in that office and for about the first time in my life I was based in one place for a while and had spare time in the evenings, so we started going out together. We did all the usual things – the cinema, pubs and meals out – and at weekends the occasional sightseeing trip inland. This made what would otherwise have been a quite lonely existence very enjoyable. On one occasion we had driven out up the valley of the River Dee toward Braemar and were returning in the dark, when we suddenly noticed that strange things were happening in the sky. Stopping on top of a hill for a better look, we were treated to an ephemeral display of green and purple wraithlike curtains writhing and twisting across the heavens. This was my first view of the northern lights, and like almost everyone who sees them for the first time I was completely enthralled.

It was 512 miles from Aberdeen back home to Croydon, which was too far to commute back and forth to see my mother every weekend, but I usually managed it about once a month, and I had a whole week off at Christmas. The boss would usually let me leave work straight after lunch on the Fridays when I was going home, and on one epic occasion the Mini Cooper S got me back in time for a pint in my local before closing time that evening. It was a noisy car and my ears didn't stop ringing for days afterwards.

Eventually the job came to an end and my next move was to the dredging investigations department at Bromborough in Cheshire, which was a subsidiary of Westminsters. Saying goodbye to the lovely Betty was a bit of a wrench, but Aberdeen is a long way from Croydon and she had other admirers anyway, so we did not stay in contact.

Dredging investigations were concerned mostly with site investigation work rather than anything much to do with the actual dredging, and the laboratory was, I think, self-financing from the money it earned doing this. Taking core samples of the ground and analysing them before a civil engineering project started was one of the main jobs, the purpose being to give the architects an idea of what sort of foundations would be required for any building or other structure. A lot of the work was repetitive and boring, and consisted mostly of putting loose samples through a series of different-sized sieves and then weighing whatever didn't drop through each one. The results were plotted on a graph. This was called particle size analysis, and is every bit as exciting as it sounds. Stuff that wouldn't go through the sieves, like the clays in some of the core samples, was subjected to a different series of tests, to see what the percentage of water was and how much load could be applied to a test piece before it squashed down or fell apart. Rock would also be subjected to strength testing, and occasionally small samples were crushed to powder and chemically analysed.

I spent a week or so in the laboratory to gain a general view of the kind of testing done there, and after that I was sent out on site with one of the drilling teams to see how the core samples were obtained. The sort of equipment used was very similar to an oil drilling rig but on a much smaller scale. At the start of the job, if the ground was soft the hollow core tubes were simply punched into the ground with a drop hammer, but as it got firmer the actual core drill had to be used – basically a steel tube with cutting teeth around the end of it. Each tube was around 6 feet long and as the drilling progressed it would sink further into the ground until there was only about a foot protruding, when the drill would be stopped and a fresh tube screwed on top, and so on. When the bore had reached the required depth, the casing tubes would be withdrawn piece by piece and numbered, so that when the cores arrived back in the laboratory they could be reassembled in the right order.

Just for a change I once got sent along with a diver to take samples from the bottom of a gravel pit somewhere near Reading, as a new road was supposed to be routed across it on a viaduct. The water was only about 40 feet deep at most, and my job was to act as second man for the diver and attend to the lifeline. Although I knew nothing about diving, all I had to do was either pay out more rope or take up the slack according to how many tugs on the rope he gave me, so it was hardly a demanding job. The diver was a really nice guy and had done a lot of work in the North Sea before deciding to semi-retire to this less arduous type of work.

He was wearing a dry suit, which consisted of a one-piece rubber outer garment which you entered via the hole at the neck, after which the rubber helmet and face mask was attached to it with a kind of sealing ring. I noticed he was wearing his normal clothes inside the suit, so he was obviously confident that it wouldn't leak. He spent no more than about 20 minutes swimming around and coming back periodically with some bottom samples, before pronouncing that he was finished.

On the car journey down there I had been plying him with various technical questions about his work, which I suppose must have given him the impression that I was an enthusiast in the making, because he now suggested that since there was plenty of air left in the cylinders I might fancy having a go. I was not averse to trying something different, so I took off my jacket and shoes and climbed into the suit. After the headgear had been attached, he got me to practise breathing through it for a minute or two before entering the water in case I had a panic attack, which apparently was not uncommon with beginners. All was well, however, and I commenced wading out into the water. After a few yards the water was so deep that my feet could barely touch the bottom, and as I seemed to be unable to sink I could progress no further. Seeing my predicament the diver gestured to me to pull open the elasticated seal at one wrist to let some air out of the suit. I did this but forgot to keep my hand above the water, which ensured that a pint or so went straight down my arm and into the suit. I sank quite well after that.

As I continued to walk out into the deeper water, I was expecting to be able to see the weed beds and maybe some fish, but my colleague had stirred up the mud from the bottom and the visibility was only a few feet, so I really had no idea where I was going. I had been told that the bottom was very uneven but as I couldn't see my feet anyway I just carried on walking. After a few more paces I suddenly dropped down a hole into some much deeper water and for a few seconds I wasn't even sure which way was up until I remembered to look where the bubbles were going when I exhaled. This gave me a bit of a scare, and for a minute or two I just stayed put to regain some composure. Following this, I decided that there was not much point in carrying on, especially in view of the very poor visibility, so I gave the required tugs on the line and made my way out again. Back on the bank, the diver asked me how long I thought I had been down for and I said about 10 minutes. He replied that actually it had not quite been 5 and that nearly all beginners overestimated the duration of their first dive. He also told me that he had guessed I had gone down the hole when the rope suddenly started to pay out faster and that I had done quite well not to panic, which made me feel a little less of an idiot. Thus ended my one and only excursion under water.

After this little job I was sent off down to Plymouth with a scientist from the laboratory to do a site investigation job in Millbrook Bay, on the western side of Plymouth Sound. We were expected to be there for about a month, and the company billeted us in a very nice family hotel in Yelverton, on the western edge of Dartmoor. The drilling crew did not work at weekends, which gave me the chance to spend some time exploring the surrounding countryside, this time in the company of one of the hotel manager's daughters; we enjoyed several very pleasant excursions to Burrator and various other local beauty spots. Up until my time with Westminster Dredging I had led a bachelor's life, being more interested in cars than women, but now I had managed to acquire two girlfriends within a few months. This one, though, tended to spread her favours, and I had the feeling that she was going out with me just to wind up one of her other suitors, so this particular romance did not get very far either.

The job itself consisted of taking core samples from the bottom of Millbrook Bay where the outfall pipes of a proposed power station would have to go. First of all, the scientist and I were going to have to set out the positions of the bore holes. Three shore marks had been set up, and we were provided with a sextant to measure the angles between them plus a special chart that we could use to plot our position. At high tides we were supposed to go out in an inflatable boat with an outboard motor, take our sights and then drop marker buoys in the appropriate places. The final positioning would be done by the drilling pontoon itself, which would place its anchors and then winch itself to the exact spot. The problem was that our inflatable kept drifting around in the wind, and by the time we had taken a pair of sights and plotted them on the chart we would be somewhere else. We had a weight on the end of a line to try and hold us in one place, but it tended to drag along the bottom and also we would swing from side to side as the wind changed.

My colleague was what most people might imagine an absent-minded professor looked like. He wore pebble-lensed spectacles, a tweed jacket with leather elbow patches, an old shirt with a slightly frayed collar, and shapeless grey flannels. He was undoubtedly very clever but somewhat lacking in common sense and he now suggested that perhaps it might be possible to simply walk out over the intended area at low water instead, so we went off for an extended lunch break while we waited for the tide to go out. Having returned to a convenient spot on the shore by mid-afternoon, we changed into our rubber boots and started to walk out in the direction of our first mark. We had not gone more than a few yards before we realised that the foreshore consisted of some very sticky grey mud of uncertain depth. It soon became obvious that we would never get out far enough in just our wellies and might even get ourselves stuck as well. The professor then came up with the idea of making some mud skis from wooden planks to help spread the weight. Sure enough, by the following day he had got hold of four planks, each about 4 inches wide by 5 feet long and some leather straps to secure them with. He was so short-sighted that he was unable to use the sextant, and the plan was that I should take the sights and he would plot them on the chart, which he had got mounted on a large clipboard. I didn't fancy trying the skis, which were simply fastened across the instep, leaving the heels free to lift up with each step, and I said that since they were his idea he should go first to prove they actually worked, in which case I would follow with the sextant.

He obviously had complete faith in his creations because he had even dispensed with the wellies and was wearing his everyday shoes instead. He set forth out across the mud with a cheery grin on his face and I confess to feeling a certain admiration for him because he was a shining example of that dying breed – the great British eccentric. At first all went well and he steadily plodded on, although I noticed that he was obviously finding it pretty difficult to unstick each foot in time for the next step. When he was out about 20 yards from the shore he stopped and looked round to call me out to join him, which is when everything went wrong. With the cessation of movement, the mud had a chance to get a better grip and when he tried to move off again he found he was stuck fast. With every attempt to lift one ski, the other was driven further down into the mud until it became obvious that he was going no further. The skis had now become a liability rather than providing any help, and he was forced to bend down and grope around in the mud to unstrap them. With the skis off, he immediately sank below his knees in the sucking ooze and floundered about for a while, seemingly unable to move. I was beginning to think that I might have to phone the Fire Brigade to get him out but eventually, by using the skis as walking poles instead he managed to drag himself back to safety.

He was pretty well plastered in mud from the thighs down and I think that most people would have made some attempt to wash the worst off, but he jumped straight in his car and went back to the hotel, where he traipsed across the carpets in the foyer leaving a trail of grey slime behind him. I was following at a discreet distance and did not hear what was said to him in reception, but later that afternoon he moved out, saying that he never really liked staying in hotels and was moving to somewhere a bit less pretentious – I daresay the hotel was not sorry to see him go either! On the following day the wind had dropped and we got the boreholes sited by using the dinghy as originally planned. In the event, we only needed to be within 10 yards with our marker buoys, as the pontoon anchors and winches could look after the rest.

This was about the last job I did with Westminster, as I had slowly come to the conclusion that as I was not Dutch and could not speak a word of the language, my promotion prospects within the organisation would never amount to much. I was also spending three or four weeks at a time away from home, and I began to think that I might as well be back at sea. I therefore started looking for another shipping job, and before long spotted an advert by Mobil Oil, which wanted engineer officers of all grades and was paying some extremely good salaries. Unlike the general cargo trade, the tanker business was booming and since I had left Esso there had been some massive pay rises in order to attract new personnel.

I applied at once and after the usual interview was offered a junior engineer's job at £1,750 per year, which in 1971, was very good money. Officers worked four- to five-month trips with two months' leave, which was also pretty good, so I accepted. My mother seemed to be managing all right on her own, and my elder brother and sister saw her at least once a week, so I hoped she would not be too lonely.

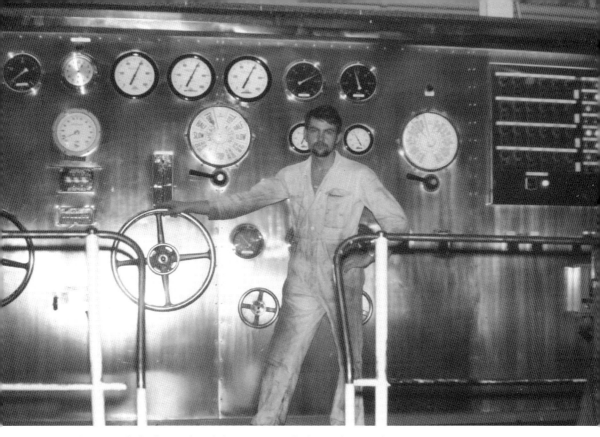

The control platform of Mobil Energy, *with the author in charge.*

Left: The author in his new uniform circa 1967.
Right: In my cabin on Esso Yorkshire.

Never judge a book by its cover – Yorkshire *turned out to be a very fine ship.*

Looking aft on the main deck of Esso Yorkshire.

Esso Durham.

On Esso Durham *in a gale in the Bay of Biscay – still hot enough down below for me to want a cold beer as soon as I came off watch.*

Mobil Daylight *loading at Kharg Island.*

Junior engineer and cadet, Esso Durham.

One of the wives from Mobil Daylight *having an elephant ride in the concrete zoo on Kharg Island.*

Engine room of Mobil Daylight – *third engineer in charge.*

Astrid on the foredeck of Mobil Astral.

Astrid beside a windlass on Mobil Astral – *note the buckled platform behind her.*

Left: Bert, a crane driver with Westminster Dredging – an unexploded bomb in the grab did not bother him!

Right: Third engineer (right) and his junior on Mobil Daylight *looking pleased with themselves.*

Left: My junior engineer (left) and cadet on Mobil Energy. *These two saved my life when I became trapped in a ventilation shaft.*

Right: 'Ed', fourth engineer, Esso Durham.

Below: Junior engineer pulling pistons out on Post Charger.

Above: Engine room rating, Mobil Energy.

Left: Fourth engineer, Post Charger, *sitting on a piston.*

Above: Mobil Energy *heading out into the Pacific from the Panama Canal.*

Below: Heavy weather – Mobil Energy *takes a thump on the nose.*

Far left: Mobil Energy, *I think.*

Left: Astrid sightseeing somewhere on Mindanao island, Philippines.

On Post Charger, *Sandakan, Borneo, I think.*

Astrid having fun crossing the Tasman Sea on Flinders Bay.

MV Tiger *in Dover Harbour. Photo courtesy of Ken Larwood.*

MV Hengist *on the beach at Folkestone.*
Photo courtesy of Nigel Thornton.

Hoverspeed Great Britain *in Dover.*
Photo courtesy of Ken Larwood.

11 MOBIL OIL

Mobil was another American company, but it ran a considerable fleet under the British flag and had an office in London, so I was still going to be sailing under the Red Duster. After formally accepting the job offer and having another medical (this time without being asked to cough!) I enjoyed a couple of weeks at home before my first orders came through: I was going to be flown out to Durban in South Africa to join *Mobil Daylight*, a supertanker of around 96,000 dwt – the biggest I had yet sailed on.

When it was time to leave home, a friend of mine from the car club, John Lane-Smith, drove me up to Heathrow. He was late picking me up and in his haste to make up the time gave me the quickest and scariest trip up through town I have ever had in a car (no M25 back then) but he dropped me off at the terminal about 15 minutes before the flight was due to leave. These days, what with all the security checks to go through, an arrival as late as that would mean missing the flight for sure, but on this occasion I simply walked up to the SAA desk, presented my ticket and strolled through the barrier as they were making the final call.

The flight out to Durban had to take an extended route mostly over the sea, because South Africa at that time was an international pariah and SAA had been refused permission to overfly many black African countries. This made it quite a long flight with nothing to see for much of the time, and I was very glad to stretch my legs when we had a brief stopover at Johannesburg. From there we continued to Durban, where I was met by the agent and conveyed to a very smart hotel in the middle of town – it had all been very efficient.

As the ship was running late I had two full days to go sightseeing in Durban before actually joining. It would have been very enjoyable had it not been for the apartheid business, which I had obviously heard of but knew nothing about in practice. It was much more extensive than I imagined, and it struck me as quite ludicrous that even the benches on the seafront were marked with signs saying what colour a person had to be before they were allowed to sit on them. In the hotel itself I never saw a black person, although I guess they may have been employed somewhere out of sight of the guests, who were without exception white. The waiters and chambermaids were coloured, and I was even rebuked by one of the other guests because I said hello to one of the maids and held a door open for her as she was struggling to push her cart of cleaning materials through it. If I ever got into conversation in a bar or restaurant with the local white residents and they found I had just come out from England, they were all at great pains to tell me how the country would go to pieces without apartheid, and how well off the blacks were compared to the way they might be living in other countries. I have never met a more bigoted and self-centred bunch of people in all my life.

After breakfast on the third morning, by which time I was quite sick of South Africa, the agent drove me down to the docks where a launch was waiting to take me out to the ship. She was lying several miles offshore because of her size, and as the sea was quite rough for a small boat I was very glad that the trip lasted no more than half an hour or I would have lost my breakfast. Although I was never seasick on a proper ship, I am no small-boat sailor, and on the odd occasion I have been out in a fishing boat I can throw up as well as any landlubber, so I was very relieved when we finally reached the ship.

I am not sure when the term 'supertanker' was first coined: in World War II the famous T2 class was the biggest around, and they were just 16,500 dwt. This remained the benchmark until the 1950s when Esso introduced a class of 26,000 dwt, quickly followed by 36,000 dwt. By the end of the 1960s ships of 100,000 dwt were quite common, after which the escalation in size was very rapid, with some ships of around half a million tons being constructed before 1980. *Mobil Daylight* at 96,000 tons might not have stacked up against these leviathans, but she still seemed ruddy big to me as we bobbed around on the swells beneath her huge slab sides, while I waited my chance to grab the rope ladder that had been lowered for me.

Having reached the main deck I was met by one of the other engineers and a steward, who helped me carry my bags to my cabin. *Daylight* was another old-style tanker with the bridge amidships and after traversing the deck we climbed up onto the flying bridge and made our way down aft – the flying bridge on this ship was some 130 yards long.

Having settled into my cabin, which was very spacious and cool, I was told that I would be put on the 12 to 4 watch and as it was by now nearly 11.30 I had better get along to the mess if I wanted any lunch. My feeling of sea sickness having disappeared the minute I set foot on the solid decks of *Daylight*, I followed my nose around the alleyway until I found the mess and sat down with one of the second mates and the third engineer, who was to be my colleague on watch.

Mobil ships employed British officers and Indian crews, so there was one item that you could guarantee would be found on the menu – curry. In fact, curry was available at every meal (including breakfast) on every day of the week, as well as the usual English dishes and I developed quite a taste for it. This basic curry reputedly came from a pot which was topped up every day with whatever the cook fancied throwing in, so it was never emptied or washed up – and pretty good the result was too. Apart from this curry *ordinaire*, a special Indian dish was served once a week for dinner instead of a roast, and this was always fabulous. We had two cooks: one prepared all the English dishes for the officers, while the second cook, or *bhandari*, as he was properly called, cooked for the crew.

The Indian crew were either recruited from Bombay or Goa – the latter having names that indicated their Portuguese heritage, such as da Silva, Fernandez, Rodriguez etc. It may seem strange, but I can assure you that there was far less of an 'us and them' attitude with the Indian crew than there had been in Esso with a British crew, and here the officer–crew relations were excellent. In Esso, the officers and crew had hardly ever mixed socially apart from on Boxing Day, when the officers would serve the men their dinner for a change. In Mobil, however, it was not uncommon for the officers to be invited to the crew's mess for a meal to celebrate some Indian festival, and these usually turned out to be very jolly occasions. The Indians' attitudes to their work also differed markedly from that of the British guys in Esso, where most of the spanner work in the engine room was done by the officers, while the ratings would simply carry on with their usual cleaning and painting. They would assist when asked, but would never volunteer, whereas in Mobil the engine room ratings would always be hovering around if there was a work-up in progress, offering to lend a hand, make the tea or just fetch and carry.

The wages they were paid were pitifully small compared to British rates of pay – I think that the average rating earned less than half what I made as a junior engineer, which is of course, why Mobil and many other shipping companies liked to employ them. By Indian standards, though, the money was excellent and there was fierce competition to get jobs on British ships. Once or twice when things were a bit tough in the engine room and the ratings were complaining, I would enquire artfully as to why they didn't join an Indian shipping company instead, at which they would look at me in horror and say, 'Oh no, Sahib, Indian company no good!' Surprisingly perhaps, the reason they preferred to be on a British ship rather than sail under their own flag was that their own Indian officers treated them so badly. This may have been something to do with the caste system, which of course had no place on our ships; it didn't matter to us whether a man was a Brahmin or what we then called an untouchable as long as he could do his job. Whatever the reason, in Mobil the officers and ratings always got on extremely well, which made for a very happy and relaxed atmosphere on board.

Going down for my first watch, I was impressed with the layout of this Japanese-built ship. For a start, the boiler control panel had been placed inside the engine room, so there was now just one huge console with the boiler controls and instruments to the left, and the engine control panel and manoeuvring wheels to the right. This meant that any irregularity in the boiler system could usually be fixed by the officer on watch without him having to go through to the boiler room. As for the plant itself, it consisted of a 28,000 hp turbine set and two Foster Wheeler boilers working at 950 psi (65 bar), superheated to 500°C. It was a well-thought-out engine room, and most things were where you expected them to be, although every ship has a few little anomalies just by way of an intelligence test for the engineers.

One thing that always struck me as potentially dangerous practice was the way that any engineer joining a ship would be expected to simply go down below and take over a watch without anything much in the way of a handover – whereas it would for example be inconceivable that an airline pilot would be asked take off in an unfamiliar aircraft without being given very thorough conversion training first. Quite often all you might get by way of an introduction to your new ship was something along the lines of: 'Beers are in the fridge and the job's down below – have fun!' This meant that the first few days aboard could be quite difficult until you found out where everything was. To avoid any problems caused by unfamiliarity, it was very unusual for a complete set of engineers to be relieved at the same time, so that a few experienced men would always remain – there was no guarantee, however, that there would be one of these on every watch. At places such as Cape Town where the crew change took place by boat, this absence of handover was unavoidable, as the launch would be having to keep up with a moving ship and could not wait indefinitely while the new boys chatted to the off-going crew – the latter, in any event, would be so pleased to be going home they might already be half-pissed by way of celebration, and therefore unable to communicate anything useful.

Returning to my cabin after my first watch, I was pleasantly surprised to find that all my unpacking had been already been done by the steward, and a fresh white shirt and shorts laid out for me on my bunk – this was service such as I had never encountered in Esso. He would wake me in the mornings at my chosen time with a cheery, 'Good morning, Sahib,' and a cup of tea – and for an extra £5 a month he would do all my laundry as well. All the stewards and most of the senior deck and engine ratings were Goanese; they spoke excellent English and were quite the most charming and friendly chaps I ever met at sea.

I soon settled back into shipboard routine on *Daylight*, which was both reliable and easy to operate. It did, however, have one quite serious problem, which after my experiences on *Durham* gave me a few palpitations: the bilges were always filling up with sea water from multiple leaks in the pipework. *Daylight* had a beam of 127 feet and if she was rolling even slightly, 40 or 50 tons of bilge water thundering across from one side to the other was a very alarming sight, especially in view of the fact that many electric pumps were installed on these lower levels and were at risk of getting splashed as the fast-moving water hit a stanchion or some other obstruction.

The source of all these leaks was the very poor-quality piping that had been used in the seawater system when the ship had been built: either it was too thin or it had been made of an inferior material. As it started to corrode, some of the resultant pitting would eventually penetrate the pipe walls. When this happened we had to try and repair the leak as quickly as possible, as it could escalate rapidly from a small pinhole into a solid water jet. When we were going round the job each watch, we would carefully scrutinise the bilges for the tell-tale spurts of water and make attempts to repair them where possible. The smaller ones could often be fixed by putting a jubilee clip around the pipe with a small piece of rubber jointing strategically positioned over the hole to form the seal. The biggest clips we had would fit a 4-inch pipe, but several could be joined together for the bigger ones. Some pipes might finish up with a dozen or more of these patches along an 8-foot length. Where possible we renewed the worst pipes, but some were too inaccessible and in any event they were mostly in use all the time we were at sea, so in the main we just lived with the problem and made sure that the bilge pump was always kept in good order.

A few times, when we had a particularly bad leak and none of the usual repairs were possible, we resorted to a technique known as cement boxing. A wooden box would be constructed around the affected area, leaving at least a 6-inch gap between the pipe and the woodwork. We were not attempting anything in the way of high-class joinery, and in fact we didn't want the fit to be too close at the bottom, otherwise the leaking water would have nowhere to escape to. Next, we mixed up a batch of high-strength, quick-setting concrete and started shovelling it into the box as fast as we could go. When I first saw this procedure, I was convinced that the water spurting from the leak would simply wash the concrete away as fast as we put it in, but this was not the case: some did get washed out at first but we just kept piling more in and as the level of cement rose in the box it gradually stopped the leak and was solid within 20 minutes.

We were bound for the Gulf again – I forget which port it was this time – after which we were going to Australia and New Zealand. When we had loaded and were on our way back across the Indian Ocean a very sad event occurred: the Bandhari failed to report to the galley one morning and was found dead in his cabin, after suffering what we presumed was a heart attack. I thought perhaps that in these circumstances he might be buried at sea, but after talking it over with the crew, the captain decided we would divert to Bombay so he could be cremated in his home port according to Hindu custom. In the meantime his body was suitably wrapped up and placed in one of the freezers alongside the meat carcasses hanging there! The *Bandhari* had been in the company for many years and was held in much affection by his crewmates, so meal times were a bit subdued until we had landed his body and his replacement came aboard.

I know that during my time on *Daylight* – and I completed two four-month trips on her – we visited Fremantle, Adelaide and Melbourne, and Whangarei on the north island of New Zealand, but I cannot recall in which order. The terminal at Adelaide was too far from town to make it worthwhile going ashore, but we did get time off in Melbourne for a bit of sightseeing, and I managed to fit in a ride on one of the famous Flinders Street trams. Adelaide and Melbourne are both on the south side of the continent, which meant regular trips back and forth across the Australian Bight to reach them. Like the Mozambique Channel, this is another stretch of water that is respected by seafarers, since it seems to get more than its fair share of heavy weather and freak waves.

On one passage across the Bight a terrifying incident took place, which I can only describe as feeling like the ship had fallen down a hole. The weather had been a little rough but no more than around force 8 or 9, which on a well-found tanker of 96,000 tons was nothing to worry about at all. We were on a ballast passage, the ship was rolling gently and I was just starting to nod off to sleep having come off watch at 4 am. I suddenly became aware that the ship was starting to angle steeply down by the bows, which was most alarming, as a ship of this size and length hardly ever pitches more than a few degrees even in very severe weather. Knowing that something must be badly wrong I was out of bed in a flash and groping for my boiler suit when there was a terrific thump and the whole ship started shaking herself violently, like a giant dog that has just come out of the water. This lasted several seconds and was followed by the roar of escaping steam from the funnel, after which the engineers' alarm sounded.

When I arrived down below it was to find everything in complete confusion: the main turbines had tripped out on overspeed and were now slowing down, the boiler safety valves were blowing furiously as the steam demand had suddenly gone from maximum to next to nothing, and the whole control console was lit up with flashing red warning lights. As if we didn't have enough on our plate we then got a phone call to say that the bridge was flooded, the radar was out and there was no steering either!

It transpired that we had been hit by an enormous wave which the second mate on the bridge had estimated was more than 80 feet high – high enough at any rate to smash the light at the top of our foremast. He had happened to glance at the radar and spotted a large blip on the screen that he thought was a rain squall approaching. It was a dark night and it was only when the wave was very close that he picked out the white crest and knew it for what it was. He said it was the most terrifying moment of his life. In front of the huge wave was an equally big trough, and it was the ship starting to slide down into this that had woken me up. Being so long and heavy, the ship had not been able to rise up again in time to ride over the wave that followed and had simply buried her bows in it like a submarine. A solid wall of water had thundered across the foredeck and smashed against the front of the bridge superstructure, breaking several of the bridge windows in the process. As the wave passed down the ship, the stern had lifted far enough for the propeller to come out of the water and start racing, which had caused the governor to operate the overspeed trip.

At first we thought we had lost all means of steering from the bridge, and an engineer went to the steering flat and operated it from there instead, getting his helm orders relayed to him by telephone. This was something that we checked periodically, but it remains the only time I ever saw it used in earnest. After some more testing we found that although the automatic pilot and the electric tiller control had been lost, the old-fashioned ship's wheel and telemotor system was still working, so we could go back to bridge steering, with a seaman on the wheel. The gyro compass had also been knocked out, so we had to use the magnetic one for a change.

Back down below it took some time to work out how to reset the overspeed trip, as nobody had ever seen this particular gadget operate before, but after about 20 minutes of furious activity we were back in some sort of order. As for the second mate and the seaman on watch – they had just had time to duck down behind the chart table as the wave struck and had escaped with no more than a soaking and a bad fright but it took a week for Sparks to dry the radar out and fix it. All things considered we had had a lucky escape, as a smaller vessel might well have been overwhelmed.

Apart from the numerous repair jobs I have mentioned in previous chapters, there was one other routinely unpleasant task that befell the 12 to 4 watchkeepers: this was soot-blowing the boiler tubes, which usually took place on the night watch. The soot-blowers were connected to a high-pressure steam line – 250 psi, I think it was – each with a row of holes in it strategically placed to send steam jets to the appropriate places. While this was going on the steam consumption was so great that the engine had to be slowed down by between 5 and 10 rpm to avoid overloading the boilers. We also needed to inform the bridge before commencing, so that they could alter course if necessary – this to ensure that the clouds of soot emanating from the funnel would hopefully be taken downwind and over the side rather than being deposited all over the decks.

The soot-blowers were rotated by air motors, with a cam to control the angle through which the steam was admitted. There were two types: the more common ones, about two dozen in total, were permanently installed at the boiler fronts and in the economiser, and took a minute or so each to complete their cycle. The second type comprised two really big lances per boiler, and they were mounted on the roof of each furnace. Apart from being rotated, these two were driven right down among the screen tubes and superheater and then withdrawn again, the whole cycle taking about three minutes per lance. It was imperative that they were watched the whole time they were in use in case they got stuck, which could lead to them being burnt out. If they did stick they had to be withdrawn manually by winding a handle on the end of the air motor shaft – a procedure that could be guaranteed to raise a sweat and give you arm ache.

From start to finish the whole procedure took about three-quarters of an hour, at least five minutes of which were spent right at the top of the boiler room, in the hottest part, opening and shutting the main steam control valves. For the rest of the time, the engineer would be having to climb up and down all over the boilers, checking the operation of each soot-blower in turn and getting slowly cooked as he did so. More than once, this little game has left me totally exhausted with my boiler suit looking like a wet dishcloth – but on the plus side it kept me fighting fit and I could eat and drink whatever I liked without my weight ever going above ten and a half stone.

Despite the best efforts of the chief steward and cooks to provide fresh fruit and vegetables wherever possible and some wonderful and varied menus, there were always a few rather sad individuals who would only ever eat the same things, day in and day out. GW on *Esso Warwickshire*, with his soup and bread roll for lunch every day had been one such, while on *Daylight* we had a chap who I shall call Curtis. His particular culinary fad was chips – never anything with them, just a bowl of chips for both lunch and dinner. I don't know what he ate for breakfast as I was on the 12 to 4 and never got up for that. Unlike GW, his diet was not supplemented by alcohol, and he was a rather taciturn individual who kept himself very much to himself.

A couple of months after joining, it became clear that he was ill: it started off with bleeding gums and a general lassitude, which some of the less tolerant put down to laziness. After a while he also developed some nasty looking blotches on his legs, while a scar on his knee from an operation years before started bleeding when he gave it a knock. As he was by now barely able to walk, he was put in the sick bay and then paid off and taken to hospital at our next port of call. He was diagnosed with scurvy, which was quickly cured by some large doses of vitamin C. His was the first recorded case of this disease in the British Merchant Navy for a great many years and he was forever afterwards known as Scurvy Curtis. I never sailed with him again, but he was

Mobil Oil

111

well known in Mobil – I am told that the only change he made to his diet after this experience was to put tomato sauce on his chips!

I was much looking forward to seeing New Zealand and the first sight of Whangarei harbour, which was totally enclosed by hills covered in lush green vegetation, did not disappoint. We were there for a couple of days, which meant that nearly everyone who wanted to go ashore got a chance to do so. On my trip into town I found the place like a ghost town, with empty streets and virtually no traffic. It turned out that the All Blacks rugby team were playing some big match and the whole town was closeted away indoors watching it on TV. Having nothing better to do, I found myself a friendly bar and did the same.

After Whangarei, the ship was going back to the Gulf again, by which time I would have completed well over the agreed four months for a trip in *Daylight*. A few others were also due leave and the company decided we would go via Singapore, which meant quite a bit of extra mileage, and do the crew change there rather than upset anyone by making them carry on past their time. That Mobil was prepared to do this shows how keen it was to keep its officers. Being an engineer on a tanker was a pretty tough job, and I think that the oil companies had finally realised that if a crew member or officer was in any way unhappy with their conditions they would simply leave after one trip and go to whoever else happened to be offering a better deal at the time.

This particular part of the voyage involved some quite interesting navigation: first off we headed up across the Pacific until we reached Cape York at the north-east corner of Australia, where we turned left through the Torres Strait, with Papua New Guinea over to starboard. The deck department certainly earned their keep for the passage through the strait, as the area is infested with reefs and many ships have been wrecked there, including *Oceanic Grandeur*, a tanker of 58,000 dwt, in 1970. From here we steered across the top of the Gulf of Carpentaria, through the Arafura and Timor Seas and finally into the Indian Ocean, where we carried on westwards for about 500 miles. The skipper then took us north through the Lombok Strait, a deep-water passage between the islands of Bali and Lombok leading into the Java Sea, before bearing away north-west up to Singapore. We had gorgeous weather the whole way and enjoyed some fine views of the islands; Mount Rinjani, a 12,000-foot not quite extinct volcano on Lombok itself, was most impressive (Wikipedia tells me it erupted in 1994).

At Singapore we were booked into the usual smart hotel for the night and had most of the following day to go sightseeing before catching an Air India flight home in the evening. This was before the days of the Boeing 747 Jumbo Jet, which had sufficient range to make the flight in one hop, and we stopped five times on the way back to London: at Kuala Lumpur, Colombo, Karachi, Rome and Paris – it seemed to take an eternity. We took off in the evening, and because we were heading west chasing the setting sun, it was still evening when we reached Kuala Lumpur and only just midnight by Colombo. Altogether I think it was about 24 hours between getting our evening meal and breakfast next morning. There were several squalling infants on board, at least one of which always seemed to be awake, which precluded any chance of sleep, so that by the time I finally got off the aircraft some 33 hours after boarding I was suffering the worst case of jet lag I have ever experienced.

Having now been in Mobil for nearly a year I had saved quite a bit of money and decided to fulfil an ambition I'd had since 1961 when I first saw an E-type Jaguar and immediately fell in love with it. The Mini Cooper S had been a lot of fun, and with my friend John Canova navigating for me we once had a run of six wins out of eight starts in club rallies. This was not achieved without cost, however, and in the two years I owned the car I had had to take the engine out no less than nine times for rebuilds and repairs. It seemed that I was spending more of my precious leave fixing the car than actually driving it, so after a week or two scouring through *Exchange & Mart* I found a car I fancied at a price I could afford. The vendor agreed to take the Cooper S in part-exchange, so for the grand sum of £350 I became the proud owner of a 1964 3.8 litre E-type Jaguar fixed-head coupé. The car had already covered some 89,000 miles and was a bit scruffy, but it was nothing I couldn't fix and I was thrilled to bits with it. In the nine years I owned it, it never broke down on the road and the engine only had to come out twice – once to do the clutch and once for a rebuild at the 100,000-mile mark.

I really enjoyed motoring around in the new car during that leave, but all too soon it came to an end and I was back to sea again, this time on *Mobil Energy*, an OBO (Oil Bulk Ore) carrier. These ships have large hatch covers and specially shaped holds so that they can carry either crude oil or solid cargoes. I was going to join *Energy* at Mina Al Ahmadi in Saudi Arabia, where we would load crude oil for an American port in the Gulf of Mexico – I think it may have been Houston or thereabouts. Having discharged this, we would clean the tanks and head up the east coast of the States to Norfolk, Virginia, to load coal for Japan, which would mean a transit of the Panama Canal and my first crossing of the Pacific Ocean. From Japan we would have a ballast passage on to the Gulf, where we would start again – this was going to be a round the world trip, in fact.

Between 1967 and 1975, the Suez Canal was closed due to the Arab–Israeli War, and so once again when we left the Gulf we had to take the long way round via Cape Town before heading up and across the Atlantic to Houston. After discharging our oil which, being done almost overnight, precluded a run ashore, we steamed off south and around the tip of Florida to reach our loading ports on the east coast, cleaning tanks en route. We called at both Norfolk and nearby Newport News to take on our cargo of coal – it was actually more like coal dust, and was intended for use in power station boilers. It was loaded by grab cranes, and by the time we had finished the whole ship was black, so that after we left port the deck crew spent our first day at sea hosing everything down.

While we were in the States the fourth engineer was sent on leave and as there were no replacements available of his rank, the company sent out another junior and I was promoted to take his place. This meant that I was now a fully-fledged watch keeper and was basically the man in charge of the engine room during the 8 to 12 watch. It was a big leap in the level of responsibility but there was a big pay rise to go with it, so I was highly delighted. Once promoted in this way, it was pretty much unheard of to be dropped back to one's original rank, so I immediately transferred all my gear to the fourth's cabin, which was a lot more comfortable than my old one and also had a forward-facing porthole, which gave me a good view out over the deck

I had always assumed that the Panama Canal must run basically from east to west to get from the Atlantic to the Pacific Ocean, but an examination of the map will show that this is not the case: at that point the Isthmus of Panama runs roughly north-east to south-west, with the canal crossing at right angles to it, so the entry from the Atlantic at Colon is actually north-west of Balboa at the Pacific end.

The start of the canal is through a flight of three locks – the Gatun Locks, which lead into Gatun Lake. One of the problems with any canal which has to climb up and over a summit, as is the case at Panama, is the supply of water at the highest level. Every time a lock is used, a lock full of water runs down to the level below, and unless it can be replenished either from natural sources or by pumping, the top levels will eventually dry up. This has caused many canals to be abandoned in the past. At Panama they are fortunate, because for some distance the canal runs through the Gatun Lake, which provides an enormous supply of water. Eventually, however, it was seen that even this would not always be enough, so in 1935 the Chagres river was dammed to create another reservoir of water for the canal during the dry season. In 2006 a multi-billion-pound Panama Canal expansion project was agreed, providing new and larger locks, with side chambers, which in theory will save 60 per cent of the water at each use, so the future of the canal is assured.

Energy was a Panamax-sized ship: one that has been built to the maximum size that will pass through the Panama Canal locks (the original locks, that is). When we entered the first lock, I was amazed at how tight a fit this actually was – it didn't look as if there was room even to drop a beer can down between the side of the ship and the lock walls. The other surprising thing was how quickly the locks filled and emptied. Suppose, for instance, you were waiting to hop ashore and the main deck was exactly level with the dock wall – if you so much as turned round for a few seconds to say goodbye to someone, you would miss your chance, as the ship would by then be too high or low to get off. I suppose the reason for this, apart from the fact that the locks have very large sluices to pass the water, is that the ship pretty well filled up the space, so there was not much water to actually shift.

To navigate our ship through the canal, no less than five pilots came aboard – one each side at bow and stern, and the chief pilot directing operations from the wheelhouse. When a ship is passing through the locks, lines are passed ashore from it to small electric locomotives called mules, which run on tracks parallel to the sides of the locks on both sides; it was these that are largely responsible for keeping the ship straight and also to provide a bit of extra pulling power to get her moving when required to do so. The whole operation was extremely slick, and we were soon on our way across Gatun Lake. Fortunately, I was off watch for the more interesting parts of the transit and was able to see quite a lot of the canal. The most impressive feature, apart from the locks themselves, was the Gaillard Cut, where the French and Americans building the canal had to carve their way through a hill at Culebra. The rock was unstable, which meant that many times more spoil had to be removed than had been originally planned, and the hill which originally stood 360 feet high finally finished up at less than half that

At the Pacific end, the canal passes firstly through the single Pedro Miguel lock and then the Miraflores locks – a flight of two. Soon after this, the ship gets to more open water and passes beneath the Bridge of the Americas, which, although now superseded by the cable-stay Centennial Bridge, then took the Pan-American Highway across the canal. This particular bridge is a fine-looking structure, somewhat reminiscent in style to the Sydney Harbour Bridge but without the sturdy concrete piers at each end, and makes a grand gateway to the Pacific Ocean. After that, we saw nothing but water for the next three weeks.

There is no doubt, that as far as the non-seafaring population is concerned the world is now a lot smaller than it was, thanks to the advent of cheap, fast air travel and people think nothing of jetting off on holiday halfway round the world to the Seychelles, Thailand and other exotic destinations. To appreciate how big the oceans really are, you just have to be on a ship and although the route from Panama to Japan is quite a busy shipping lane, every time I came up from the engine room and spent a few minutes scanning the horizon while downing a beer, there was nothing to be seen. On our passage diagonally across the Atlantic from Cape Town it had been the same, apart from the rather bizarre sighting of a three-masted sailing ship which we spotted on the horizon one evening, proceeding serenely on her way like some latter-day *Flying Dutchman*.

Just to break the monotony of three weeks' continuous watchkeeping at sea, we had one blackout en route and it had a most unusual cause. A couple of weeks after leaving Panama we had a quite prolonged spell of heavy weather – nothing spectacular, but with seas big enough to sweep the main deck at regular intervals. One of the donkey-man's regular jobs was to dip the fuel tanks every morning, and unfortunately, on the last visit he made before the bad weather prevented further excursions onto the deck, he had forgotten to close the ullage cap on the tank lid. This meant that there was now a 4-inch opening into the tank, and every time a big wave washed over the deck, a hefty dollop of sea water found its way into the fuel.

On a motor ship all the fuel is centrifuged before use to get rid of any water, but on steamships there are no precision injectors and fuel pumps to be damaged, so simple filtration is all that is required. The only way the watch keeper would realise there was water in the fuel would be to try the manual drain valves at the bottom of the settling and service tanks, but on *Energy* this only got done once a watch and on this occasion it just wasn't enough.

I had just come down below and was chatting to the off-going junior, when the fireman came running in from the boiler room, shouting, 'Sahib, Sahib, come quick, boiler fires going out!' Sure enough, when we got in there the brilliant flames that could normally be seen through the mauve glass spyholes were now just a few intermittent flickers, and those died completely even as we watched. The steam pressure was falling fast and the fuel pressure had gone sky-high to try and compensate for it, so we had to rapidly change things over to handomatic while we sorted out the mess. The third engineer, who must have seen a similar situation once before, immediately twigged that there must have been water in the fuel, and sent me back to the control platform to shut the main engine throttle to save steam and to sound the engineers' alarm to bring down some reinforcements.. Meanwhile we started draining water off from the tanks as fast as we could, and as soon as there were men to spare they started at the burner fuel pipes, slacking back unions and bleeding the lines through.

By the time we eventually started to get oil through to the burners the steam pressure was getting dangerously low, while the engine room was becoming increasingly gloomy as the genny slowed down. Finally we were ready to light the torch and flash the boilers up again, but in our haste someone dropped the matches in the tray beneath the burners, which was by now swimming in water. Despite the fact that nearly all the engineers were smokers, it turned out that we didn't have a light of any sort between us, so the cadet was sent up post haste to get a box from the mess. By the time he got back down again it was just too late and we blacked out – all for the want of a box of matches! Mobil ships were much better prepared for restarting the main plant than Esso's and I seem to recall that we got going again without any of the horrible struggles I had experienced on *Edinburgh* and *Durham*.

After this excitement, we continued on our steady way to Japan – I am fairly sure that our first port of call was Chiba, at the top end of Tokyo Bay. What I do remember is that when I came off watch sometime after 4 am one day, and just before we entered the bay proper, I went to the bridge to check the ETA and caught a glimpse of Mount Fuji (or at least the top of it), just catching the first rays of the sunrise. Even from a range of some 40 miles it was a beautiful and ethereal sight. Although I made several subsequent trips to the ports around Tokyo Bay I never saw it again – probably because of the very poor air quality at the time in this, the most built-up region of Japan.

When we had tied up some hours later, the hatches were opened and we started unloading our cargo of pulverised coal, which was no doubt destined for some power station. Shore cranes with big grabs were employed on this duty until the hold was emptied so far that it was difficult to get a full load in the grab. At this stage, work stopped for a few minutes while a rubber-tracked bulldozer was lowered into the hold. This was used to push the remaining coal within range of the grab. When it became difficult even for the bulldozer to provide a decent load, men with shovels and brooms took over the job and loaded the last few hundredweight into the grab. Conditions were bad enough on deck, with all the coal dust blowing around – what it must have been like for the men working in the hold I can't imagine.

While we had been in the States, I had done a watch for the third engineer, who had a relative he wanted to see in Houston, and now he offered to return the favour so that I would have time for a decent run ashore myself. Dave C, the junior second – an amiable giant from Yorkshire who stood around 6 feet 4 inches tall – was going as well, and asked me if I wanted to go with him to save on taxi fares. This seemed a good idea as he had been to Japan several times before, so we both went ashore together after breakfast. He certainly seemed to know his way around this particular port and even had a few words of Japanese with which to direct the taxi driver. After a somewhat scary ride into town, we soon found ourselves ensconced in a comfortable bar, which

IN tHE tREAcLE MINE

116

Dave obviously knew was going to be open at this early hour. We settled down to enjoy a few beers, whilst observing the passing pedestrians through the window. The women I noticed were split about 50/50 between those wearing traditional Japanese kimonos, complete with blocked shoes and that strange cushion thing fixed to their lower back, whose function I am still unsure of. The remainder were all wearing western dress, with quite a few mini-skirts to be seen.

We were the only Europeans in this particular bar, while the rest of the clientele appeared to be businessmen. They seemed to be a very melancholy bunch, hardly ever speaking except to order more drinks – invariably whisky. These they knocked back each time in a single gulp as if they were taking an unpleasant medicine – after sampling some of the local brands, I could see why. This was of course, many years ago and since then I am told that the quality of Japanese whisky is much improved.

We spent about an hour or so in this bar over a couple of Suntory beers, which we didn't think much of either, having been drinking Tennent's for the previous couple of months. As things were very quiet, Dave then suggested we move on to another establishment he was familiar with; this turned out to be a strip club and was a short taxi ride away. The driver obviously thought he might get away with ripping us off by going the long way round but a few sharp words in Japanese from Dave soon had him on the right track. As I said, Dave was normally a very easy-going individual but he was at least a foot taller than the driver and could look pretty tough when roused, so we were soon deposited at our destination with a lot of bowing from the driver.

The club was a single-storey wooden building with a stage and a catwalk extending back into the audience, rather like a fashion modelling salon. I had no idea at the time what exactly Dave had taken me to because there were no billboards outside displaying any nude models, and I thought it might actually be a Japanese theatre. To confirm this idea, when the models appeared at first they were all wearing traditional dress with not so much as a bare ankle in sight. When the show started, there was a lot of ritual shuffling around and twirling of fans in time with some Japanese music that I found quite excruciating, but not much else. This went on for around a quarter of an hour, when quite abruptly the music stopped, all the cast lined up across the stage and everyone in the audience clapped politely. I looked across at Dave and enquired, "Is that it, then?" to which he replied, "No. Just wait and see."

After everyone had finished clapping the music restarted – this time it was much more upbeat but still pretty awful, while the cast quite suddenly and without ceremony took all their clothes off. If this was supposed to be striptease, it would appear that they didn't think the tease bit was worth bothering with. I noticed that there was very little clothing worn beneath the traditional exterior garb – presumably to make it that much quicker to get undressed.

Now that they were all naked as nature intended, the girls proceeded to take turns parading up and down the catwalk so that everyone could get a good look. One or two of the clientele, who may well have been regulars and were sitting in the seats beside the catwalk, had a particularly interesting view, because certain girls would squat down right in front of them, displaying everything they had to offer at a range of around 2 feet, which invariably earned them an individual approving clap!

When the show was over, we went outside again and hailed a taxi to take us back to the ship. This time the taxi driver really did manage to get us lost, and nothing Dave managed to say appeared to make any difference, so that by the time we finally got dropped off at the gangplank, it was some 20 minutes after shore leave had officially ended. The chief was pacing up and down the deck waiting for us, as the ship had finished discharging early and was already on stand-by waiting to sail. He started giving us the expected bollocking for being late but Dave was quite unperturbed by his ill humour and simply waited for him to draw breath before replying 'Keep your hair on, Chief, at least we're back now, so you won't have to get your hands dirty down below!' This was, I thought, rather more insolence than I would have been prepared to take, but the chief just stood there looking exasperated. In truth, he was yet another absentee chief, and the ship could have quite happily carried on without him – which would not have been the case if Dave and I had actually missed the ship.

After Chiba, where we had discharged only part of our cargo, it was off to another Japanese port to get rid of the remainder, following which we returned, as planned, to the Gulf via Singapore and the Malacca Straits, thereby completing the circumnavigation of the world in three months. Not quite *Around the World in 80 Days*, but close. On our next arrival back in the States my time was up, too, and two weeks later I was paid off in Balboa.

My leave passed all too quickly, as usual, and the expected letter came through the door with joining instructions and airline ticket for my next voyage. It was going to be *Mobil Energy* once again, and I was to catch a flight to Philadelphia and wait there until the ship arrived at Paulsboro refinery, some 15 miles south on the Delaware river. Paulsboro itself is a very small township in New Jersey, and the refineries are the main reason for its existence. It does have one claim to fame, in that it is situated on the site of Fort Billingsport, the very first piece of property acquired by the United States on the day following the signing of the Declaration of Independence in 1776. The fort is no longer there, but the land on which it was built is now a public park of the same name.

Mobil ships at that time did not carry electrical officers, and the responsibility for all the electrical work was allocated either to the third or fourth engineers – in this case it was me that drew the short straw. The chief engineer on a ship is responsible for all the machinery on board, from the main engine down to the food mixer in the galley, and part of the seconds' and chiefs' examinations included a paper on the electrical plant, so we couldn't say that it wasn't part of our job. Nevertheless, most engineers at the time tended to avoid having anything to do with the ship's electrics if at all possible, and there was usually a lecky to do the work but in Mobil it was unavoidable, and on balance I think this made better engineers out of us. Fortunately this was in the pre-silicon chip era, so that although there was all the usual generating equipment, switchboards and motors etc, the radars and some of the other navigation equipment were the only things that had really complex electronics – and these were looked after by the radio officer, so we didn't have to worry about them.

I was put on the 12 to 4 watch and was allocated a cadet as well, so there were three of us to share the work. I delegated all the routine lamping up (replacing blown light bulbs and neon tubes) to the cadet, which meant that I didn't have to do field days just for this rather boring task. I don't know whether it was the vibration of the ship that caused so many bulbs to blow, but it was amazing how many needed changing – at least 50 bulbs and tubes a week, I would guess.

The bulk of the electrical work did in fact consist of simple stuff like the aforementioned lamping up, greasing all the motor bearings and fixing the odd toaster or vacuum cleaner for the stewards. There was one other rotten job, however, that occurred several times on *Energy* and which in the end nearly cost me my life. The ventilation fans at the top of the boiler room consisted of motor and fan units mounted inside their respective trunkings, which supplied air to the working regions lower down. The trunking was split in two, with the motor bolted onto the moving section, so that it could be swung out for access for maintenance. The problem was that the flexible cable that connected the motors on the moving sections to the junction boxes on the outside of the fixed sections was of rather poor quality, and over the lifetime of the ship its insulation properties had broken down, necessitating replacement. As I said before, all these fan units were right at the top of the boiler room where the temperature was usually 130°F or more, which made it a ghastly place to have to work for any protracted period.

I used to do this job on the 12 to 4 am morning watch, as this was the quietest time, and possibly a tad cooler than the afternoon one. The first time I had to do it, I found that around ten minutes was the longest I could stand at a time without descending to the engine room for a blow and a drink at the water cooler. After several sessions like this, I had disconnected the old cable at the motor end and then had to feed it back through the hole at the back of the junction box to remove it. While doing this I noticed that there was quite a decent natural draught coming down the trunking which was refreshingly cool and that furthermore, there was a metal grid a few feet down – presumably to catch any debris should the fan shed a blade, which looked solid enough to stand on. I tentatively climbed inside the trunking and while I could still hang on I slowly lowered my weight onto the grid. It seemed pretty firm, with hardly any give in it. This meant that I could now stand and work inside the trunking, which was a whole lot cooler and I could get the job done that much faster.

Over the next week I did three of these fan jobs – but on the fourth and last one, the grid, which must have been a bit more corroded than on the previous three, gave way and I fell through. There was nothing around me to grab, but I had the presence of mind to stick my elbows out, which stopped me from disappearing completely down the trunking and out of sight. Around 15 feet lower down, the trunking split in two and I had horrible visions of one leg going down each side, which would leave me completely stuck and helpless. It would be unlikely anybody would hear my yells above the general noise in the boiler room and I might well die before being found, as it was still ruddy hot in there even with the slight natural draught.

I don't know how long I spent in there like that, with my legs hanging in space – absolutely no purchase for my feet and the broken edges of the grid sticking in my chest – but it seemed like an eternity. I was yelling and shouting the whole time, but there was nobody there to hear me. Any attempt to lever myself upwards simply drove the sharp edges even further into my torso, and I soon decided that the only thing to do was to try and conserve my strength and await rescue. I was terrified that I might pass out, which would inevitably mean that my arms would relax and allow me to fall right through.

Luckily for me, my junior on the watch was quite a bright lad and when I had not put in an appearance at the water cooler for half an hour, he decided to come looking for me – I can't put into words how relieved I was to see his face appearing over the edge of the trunking. I yelled at him to go and get a rope from the workshop, which he quickly did, and then tied it under my arms and around my chest. With him and the cadet, who had also turned up, taking my weight, I was at last able to use my hands to free the grid from around my body and wriggle upwards and out. I had suffered numerous cuts and scratches right around my chest and back but was otherwise unhurt. A very lucky escape.

I had one other unpleasant experience while doing lecky work on *Energy*, which happened when I was doing nothing riskier than changing fuses. All the starter boxes for the main motors on the ship used the old-type screw-in porcelain fuse holders with a little glass peephole at the end through which the tell-tale colour-coded disc of the cartridge fuse inside could be seen. On this occasion, we were about to strip down a compressor for repair and it was always my practice not just to open the circuit breaker while such work was in progress but also to remove the fuses and keep them in my custody, to ensure that it would be impossible for anyone to start the machine by mistake. I duly went over to the starter box to remove the fuses. The first two came out without any trouble, but on the last one I failed to notice that a chunk of the porcelain at the bottom of the fuse holder had been broken off, exposing the bare metal beneath. As soon as I gripped the fuse I got an almighty belt of 440 volts, which actually blew me back a few yards across the plates. I was not burnt, but my arm was completely numbed and felt like it had been put through a mangle – it was several hours before it returned to something like normal. I would have dearly liked to meet whoever had put that broken fuse back, so that I could kick his backside into orbit!

I had now completed my four-month trip and the ship was approaching the Panama Canal for the second time since I had joined in Paulsboro. I was due to be relieved but as no telegram had arrived I guessed that I would have to do another three-week Pacific crossing and then fly back from Japan – but events at home about which I knew nothing at the time would cause a change of plan.

During our transit of the canal, I was on duty at the manoeuvring wheels as we were passing through the first set of locks. I enjoyed being the engine driver, as there was rather more to the job than simply turning one wheel to go ahead and the other to go astern – in fact an unskilled man at the wheels can get a steamship into all sorts of trouble. The problem is that in a water-tube boiler, as I have explained before, there are an awful lot of tubes which will be filled with steam bubbles as the water boils. When steam demand is very high – when going full astern for instance – the water will be boiling so fast that there will be a lot more steam bubbles than actual water in the tubes. If the next movement is Stop Engines, the steam demand will be cut by around 80 per cent, so most of the fires in the furnace will need to be extinguished to avoid blowing the safety valves. In turn, the water will no longer be boiling nearly so fast and most of the steam bubbles will collapse – this will cause the water level in the boiler to drop like a stone, as the feed pump will be unable to refill it fast enough. If the level gets too low, the low-level trip will operate, which will kill all the fires!

The remedy for this is not to be in too much of a hurry to close the astern wheel, to give the boiler room crew time to pull the burners out and then to snap the wheel closed and open the ahead wheel to use steam there instead, to brake the shaft to a stand. Usually, in order to make up for this delayed response, I would even overshoot slightly, and finish up doing perhaps 10 rpm ahead, so that the astern turbine could be a given another shot of steam, to use some more up. In any case if the bridge didn't feel the ship was slowing down fast enough, they would ring down a movement in the opposite direction anyway, which we would be more than happy to give them. Good manoeuvring also needs a good man at the boiler console to increase feed pump pressure in anticipation of any sudden drops in water level such as I have described, and also to carefully regulate fuel and air pressures to maintain good combustion without making too much black smoke – a real team effort, in fact.

After we had cleared the locks and were sailing across Gatun Lake, I was surprised to see the third engineer come down the ladders an hour before time to relieve me. All he would say was that the captain wanted to see me on the bridge. I wasn't sure what misdemeanour I might have committed to require such a summons, but I did not have long to wait before finding out. As soon as I got to the bridge the captain took me aside and said that he was sorry to tell me that a radio message had been received via our London office to say that my mother had been taken seriously ill; arrangements had already been made for me to be paid off in Balboa. The captain had got the articles out for me to sign off, and then gave me back my discharge book. There was not much time for me to do anything except to get washed and changed, pack up all my stuff and pay the steward for doing my laundry and generally looking after me.

The ship did not stop at Balboa, so I went ashore in the pilot boat, where I was met by the agent on the quay, who had arranged a taxi to take me straight to the airport. The airport was at the other end of the canal, and it seemed an interminable drive through the jungle on some not very good roads to get there. There was not much time before the flight, and I tried in vain to find a telephone that worked for international calls so as to get in touch with my sister and find out what had happened. The first leg of the journey was to Miami, where I had to change planes. This caused some problems as my American visa was out of date and the concept of transit lounges did not appear to have taken hold in the States at the time. Even though I had no intention of setting foot on US soil, this cut no ice with the immigration authorities, who allocated an armed escort to stay with me at all times to ensure I didn't try to do a runner. At least in Miami I found a phone that worked – but there was no reply either at home or my sister's house, and it was not until I got back to Heathrow that I was able to speak to my sister Jean, only to be told me that Mum had died in hospital the previous day.

The following week flashed by, what with the funeral and all the legal procedures to be dealt with, but after this I was left to wonder about what to do next. Under the terms of my parents' will I inherited one-third of their house in Shirley, and as I was living there anyway at the time I did not have to worry about finding a roof to put over my head in the immediate future. My brother and sister were considerably older than me and were both married and living in their own houses; they very generously did not put me under any pressure to sell the house to get their share, or to raise a mortgage and buy them out, although the latter is what eventually happened.

I had by now accrued enough watchkeeping experience with Mobil to be entitled to sit for my second-class Dept. of Transport engineer's certificate, for which they would be obliged to pay me study leave for at least three months. This, together with the two months' ordinary leave I was owed, meant that I would not have to go back to sea for five months. The complete examination consisted of several theory papers, including thermodynamics, maths, mechanics, naval architecture, electrotechnology and engineering drawing, in addition to two practical papers on general engineering knowledge and either steam or motor engineering, depending on which sort of ships the engineer had served on – in my case, it would of course be steam.

By virtue of my OND I already had a head start on those engineers who had served their time in a conventional apprenticeship, so I was exempted from the maths, mechanics, heat engines and drawing papers. Nevertheless, I had to do the naval architecture and electrotechnology papers again. So my exam would be four three-hour papers plus an oral exam afterwards in front of a panel of ex-marine engineers. This latter exam was for many the most nerve-racking part of the whole process because you could be asked about absolutely anything in the ship's equipment, from the main engines and boilers at one end of the spectrum down to the fridges and air conditioning at the other.

Before being allowed to sit the exams, all candidates (deck and engine alike) were required to have taken and passed a Merchant Navy firefighting course. The one I was sent to was held at the MacDonald Road Fire Station in Leith, and was reputed to be the toughest such course in the country. Indeed, when we were being given our introductory talk by the chief instructor, he informed us with a sadistic grin on his face that some of us could expect to get burnt – but only slightly.

The first couple of days were quite innocuous, consisting of classroom work, breathing apparatus instruction and some practice with handheld extinguishers out in the yard, where small fires of various types had been lit. The most impressive demonstration was what happened if water was used to extinguish a chip pan fire. The pan had been allowed to burn for several minutes beforehand to ensure the oil was really hot, and then half a cup of water was tipped into it from a ladle on the end of a long handle. This caused an amazing fireball to erupt from the pan, with the flames reaching about 30 feet into the air. Our gallant instructor then proceeded to put it out by simply using a fire blanket – an impressive feat, we thought.

The real fun and games started on Day 3, when we were expected to put out a fire in a mock-up of a ship's engine room that had been specially constructed on the site. This creation had been made of a number of steel shipping containers welded together, and stood several decks high. It had some internal partitions and proper engine room-style ladders going up and down inside it. For our main exercise, we had to go in at the top wearing breathing apparatus and try and put out a fire that had been lit by the instructors at the bottom.

Our weapons to fight the fire were the standard 2.5-inch hoses and nozzles with which most ships were equipped at that time. The difficulty of handling these hoses when charged with water at firemain pressure is something that has to be experienced before it can be believed – even with a three-man team it took all our combined strength to manoeuvre the wretched things down the ladders and around corners to actually reach the seat of the fire, while our gleeful instructors were scurrying around at the bottom pouring more diesel on the flames to ensure realism. Of course, hot air rises, so that the temperatures at the top of the structure reminded me of *Durham* on a bad day, but at least I was prepared for it, unlike some of the other candidates. One of them hadn't been inside the chamber for more than a few seconds before stepping right back out again. No amount of persuasion could get him in for another go, so he was scratched from the course and presumably would not have been allowed to take his certificate. I felt quite sorry for him and for some of the other chaps too – I was young, fit and used to hot engine rooms, while some of the others were considerably older and were expected to handle temperatures that might have been quite outside their previous experience.

We had all been issued with firemen's jackets, trousers and gloves, but I was last in line to collect mine and the only jacket available was several sizes too small, which meant there was a gap between where the sleeves ended and the gloves began. I had not been inside the 'engine room' for more than a minute before the exposed area of my arm brushed against a handrail so hot that it simply wiped a strip of skin clean off – our instructor hadn't been joking when he said we might get burnt.

Eventually our team managed to get into position with our hose and we could actually fight the fire. Although it was an oil fire burning over an area of at least 20 by 10 feet, it was quite easy to put out with water – provided we remembered to use the spray setting on the nozzle and not the jet. When the fire was finally killed, we were all nearly out of air on our BA sets and were very happy to be allowed to use a door at the bottom of the chamber to get out into the fresh air again. All in all, I think the instructors had done a grand job of showing us how difficult it would be to fight a fire under real conditions, when we had struggled so hard with the simulated ones. We all agreed that the 2.5-inch hoses were real pigs to handle, and the instructors told us that they had been campaigning for years to get the authorities to change to 1.5 inches instead for all internal firefighting on ships – the ease and speed of handling of these smaller hoses in their opinion more than cancelled out the reduced quantity of water they could supply.

The study courses for the exams were held at various seaports up and down the country, and for me it would mean another spell at Poplar Tech, which I was not exactly looking forward to. Mr Burrage, who was to take us for naval architecture, was still there, and was his usual jovial and friendly self. He even remembered me from the model ship incident and asked me whether I had sunk any more ships by counter-flooding the wrong end, which gave us both a laugh. So was Dr S, and he appeared to be as grumpy as ever, although fortunately I did not have to attend any of his classes. As we were no longer cadets, we couldn't be ordered about like schoolchildren any more, and most of the classes were enlivened by some amusing banter between students and lecturers, which helped the process along.

Once again, I was faced with the dreary commute up and down to Poplar, and the traffic was even worse than I remembered from before, so I seldom took the car. One other difference I noticed was that the number of ships in the West India Docks at the back of the college had dwindled alarmingly: whereas in 1964 there had been a forest of masts and funnels appearing over the warehouse roofs, now, in 1973, there were only a handful of ships to be seen.

The exams themselves were held every month and you didn't have to go through the full college course before having a go. A few of us decided to have a crack after just the first two months – if we passed, all well and good, and if we failed it wouldn't matter too much and would also show us what we needed to brush up on for the next attempt. Only if you failed three times in a row would you be sent back to sea.

The two practical papers were, I thought, not too bad, and I also felt I had done enough to pass in naval architecture, but the electrotechnology paper was a swine and had me really struggling with the calculation work. Fortunately I had been schooled rigorously by Dr S, of all people, in the importance of good exam technique – making sure you attempted the required number of questions, even if you could only answer a small part of some of them, rather than spending more time on the ones you could do. It was also important to ensure that you showed every step of your calculations; that way even if you did the arithmetic wrongly, the person marking it could see whether or not you were on the right lines. Anyway, I had given it my best shot and then had to wait a nail-biting day or two for the oral exam.

It was usually reckoned that if this lasted between 20 minutes and half an hour then you would definitely pass, three-quarters of an hour was average, while if it got to an hour in front of the panel then it was probably going to be bad news. I can't remember very much about mine, except that they managed to find plenty of things to ask me that were completely outside my experience to date and about which I had to think very hard before improvising an answer – which was probably their intention. All in all, it was turning into an extremely uncomfortable interview, so I was very much taken by surprise when they stopped after about 20 minutes, conferred amongst themselves briefly and then said 'Congratulations, Mr Richardson, you have passed your second-class certificate!' I have to say that I was both amazed and delighted and walked out into the dreary East End streets with a real spring in my step.

Now I had my second's ticket, I knew that I would be guaranteed a third's job in Mobil on my next trip, with a quite substantial pay rise to go with it. It was an attractive prospect, but there was a problem in that it would mean leaving the house empty for four or five months at a time whenever I went off to sea. Even though it was situated in quite a decent neighbourhood, I still felt that the uncut grass and other signs of neglect would eventually attract burglars or squatters, so I decided to take some time out from Mobil while I worked out exactly what to do. They were very decent about it when I handed in my notice and said that they would be very happy to take me back any time I changed my mind, which I thought was a pretty generous offer. Meanwhile,

I took a job with a firm of toolmakers in Croydon, which paid a reasonable hourly rate and also increased my fitting and machining skills considerably.

Shortly after this, I was driving home one day in the E-type when I chanced to stop at a side turning to let a car out in front of me – I was surprised to see that it was being driven by the mother of one of my old school friends, who also happened to be a good friend of my late mum, so we exchanged a friendly wave. Much more interestingly, sitting alongside her was what to me looked like the most beautiful girl in the world. I knew this must be her daughter, Astrid, who had been just a little girl with long blonde plaits the last time I remembered seeing her. In the meantime, the miracle of growing up had taken place and she was now quite stunning. I was even more surprised a few days later when they paid me a call at home – they had by now heard about my bereavement and had come to see how I was getting on. To cut a long story short, I started dating Astrid afterwards, and much to my delight and amazement (for I had never had much luck with the opposite sex beforehand), she accepted my proposal of marriage after just a three-month courtship. A few more months after that we got married – and remain so some 47 years later.

After a wonderful honeymoon in Switzerland, which just about bankrupted us, we enjoyed a few months of married life at home before the lure of earning some really good money finally persuaded me to go back to sea again. As Mobil had been pretty decent to me, I approached them first and they seemed only too pleased to have me back, this time as third engineer, which meant another big jump up the pay scale. Within a month both Astrid and I were flying out to Rotterdam to join SS *Mobil Astral*, a VLCC of around 150,000 tons – the biggest ship I was ever to sail in. Tanker companies were finding it hard to keep personnel at the time, and apart from paying very good salaries, many of them allowed all the officers to take their wives with them – a privilege that had previously only been available to captains and chief engineers.

There were no worries about who was going to look after the house now, as Astrid's parents, who lived in a small maisonette without a garden, were only too pleased to be able to come down and mind it for us – in fact, I told them to treat it as their own house while we were away. Astrid's dad was a keen gardener, and although they also had an allotment I think he really enjoyed pottering around in our garden.

Astral was actually a sister ship of *Daylight* but had undergone a process known as jumboisation, in which a ship is cut in half and an extra section is welded in, to increase the carrying capacity – in the case of *Astral* by about 50 per cent. At the time, this made her one of the longest ships in the world, and almost certainly the biggest tanker to retain the amidships bridge structure. The great length of the ship meant that under certain sea conditions, the hull would flex quite considerably. If you were down aft and looked down the side toward the bows, you could actually see the movement, as the ends went up and down a couple of feet or more relative to the amidships accommodation – I hoped that the man who had done the longitudinal strength calculations had done his sums properly. The flying bridge was over 150 yards long and in really bad weather on a loaded passage with the ship deep in the water it was quite risky to attempt to get from the bridge down aft and vice versa. To bear witness to the hazard, one of the steel bus shelters was leaning over at least 30 degrees to the vertical, where it had once been struck by a particularly big wave that had swept the deck. The captain told me that on another occasion, one of the stewards had attempted to bring him his hot dinner up the flying bridge in heavy weather, using a plate with a metal lid over it. He had been forced to take cover in a bus shelter as he spotted a big wave crashing over the rail but it had roared in through the open ends and soaked him to the skin anyway. When he arrived, dripping water all over the captain's carpet, and took the lid off, the dinner had gone missing and all the plate contained was half a cup of cold sea water!

Tankers are not cruise ships and anyone thinking that working on one might be a good way of seeing the world and getting paid for it would be severely disappointed; in fact, this voyage on *Astral* was one of the most boring I ever did. After leaving Rotterdam, we sailed to the Gulf, with only the usual mail stop at Cape Town to break the monotony. Our loading port turned out to Kharg Island again, so Astrid's first run ashore was the same as mine had been – a walk down the jetty to visit the seaman's club and the concrete zoo.

From Kharg it was all the way back round the Cape of Good Hope and up toward Europe, only this time we turned right into the Mediterranean and finished up at Trieste in northern Italy, where we did finally enjoy a day ashore. This particular leg of the voyage was enlivened when we broke down somewhere off the west African coast – condenser trouble again, I seem to recall – which meant we spent half a day stopped and drifting while repairs were effected.

It was a lovely sunny day with a calm sea of a beautiful translucent blue, which looked just right for a swim. I was standing on the poop after my watch, enjoying a beer and chatting with my junior while we stared idly down into the water, when I suddenly spotted a dark shape pass by, almost right under the stern. We both watched quite intently now and after a minute or so it appeared again, this time near enough to the surface for us to identify it as a largish shark which was soon joined by a couple of his mates – just as on *Warwickshire*, this quickly dispelled any thoughts of going in for a dip. Once again, the nobby clarks had found our ship in the vastness of the ocean within a couple of hours of it stopping.

We were not the only ones to have spotted the sharks, and some of the crew had gathered at the rail and were gesturing and chatting excitedly. There were a number of very keen fishermen among them, who would always have a line over the side whenever we were in harbour or simply stopped at sea, as in this case. For the most part they only ever caught small fish, which they would invariably identify as snappers when asked what species they were.

Now that there was some bigger prey to be had, the tackle was soon changed for the heavy duty version, which consisted of a large steel hook concealed in a small hessian sack filled with offal and meat offcuts. This was tied to a strong steel trace wire and then to some braided polypropylene line of about 150 lbs breaking strain. The bait was suspended some 10 feet beneath an empty 5-gallon cooking oil drum, which acted as a float and would be a drag on a running fish. The baited hook was allowed to drift astern until it was about 25 yards away, and then the waiting game commenced.

The sharks noticed the bait almost immediately but for some 20 minutes they simply circled round without showing much interest. Eventually, though, one of them made a lunge at the bait and the oil drum slid rapidly across the surface for a few yards before submerging and disappearing from view. The intrepid angler continued to pay out line for several more seconds before pulling the line up tight to drive the hook home. I did not think for one minute that the line would be strong enough to hold the fish and that it would soon be broken, but I had reckoned without the skill and experience of this particular fisherman. Having ensured the shark was hooked, he did nothing to stop it when it swam away but simply paid out line from a carefully laid-out coil on the deck. This first run took out 100 or so yards of line, but eventually the shark, which so far had not felt any resistance beyond the steady drag from the oil drum, turned and started to swim back again, which enabled the line to be retrieved. Now a little pressure was applied, which started another run, and again line was simply paid out until the shark turned back. This procedure went on for about 20 minutes, by which time the runs were becoming noticeably shorter and eventually the shark was mostly swimming back and forth across the stern of the ship, sometimes thrashing the surface, which no doubt used up more of its strength.

By this time, nearly all the off-duty crew were lining the rail watching the proceedings and a few bets were being made as to whether or not the fish would be landed. Eventually the shark tired enough for it to be drawn alongside and now I wondered how on earth it could be brought aboard, as the line was quite obviously not strong enough on its own. At this point the captain appeared armed with his revolver (to quell mutiny, I wondered,) which was normally kept in the ship's safe. He had a rope ladder rigged over the side and then climbed down, gun in hand, to administer the *coup de grace*. Three shots to the head were sufficient to stop any further struggles, after which a stronger line was tied around the tail and the shark hauled up using the mooring winch – not very sporting perhaps, but the only way possible under the circumstances.

When you consider that big game fisherman usually strap themselves into fighting chairs on the back of their boats and employ powerful rods and complex reels with variable drag settings, our crew member's performance with just a few hundred yards of handheld line was quite remarkable – I am not sure that he even had a pair of gloves!

Measured up, the shark was nearly 9 feet long and had white tips to the dorsal and tail fins, which pretty much identified it as one of the oceanic white-tipped variety. I will not go into the gruesome details of how this noble creature was butchered, except to say that most of the carcass was thrown back into the sea, where it no doubt provided a meal for its former colleagues. The crew were only interested in the fins and the offal, although I believe the jaws and teeth were also extracted as a trophy by its captor. Within an hour or so the deck had been washed down and no trace of the poor beast remained apart from the intestines, which had been strung out on a line to dry in the sun and finally resembled nothing more than ragged brown strips of chamois leather, after which they were apparently fit to be eaten. Neither Astrid nor I fancied it much.

Several months at sea with little to do apart from reading, sunbathing and drinking in the bar could easily become a recipe for boredom for the officers' wives on board ship, who of course did not have any sort of job to take up their time, and one or two never really settled down to the shipboard routine. For the most part, the resulting trouble confined itself to a little drunken flirting, although on one occasion there were actual fisticuffs between two of them – they had somehow managed to fall out over who should have first choice from the fridge when making sandwiches in the officers' pantry one night. Fortunately, Astrid had started a correspondence course to study for a history degree, so whenever I was down on watch she was working at her studies, which filled in her time nicely. After three years of this she sat and passed a London University external degree, which meant she was well qualified to advance her teaching career when we eventually gave up deep sea voyaging.

After discharging our cargo in Trieste, where we finally managed to have a decent run ashore, it was back around the Cape again for another visit to the Gulf, this time to Umm Said, where the size of our ship precluded us from actually tying up to the mainland and we loaded from an artificial platform some way offshore. We had to wait a couple of days for a berth, and the captain agreed that some of us could take a lifeboat and go sightseeing. The sights consisted of a small sandy island which was not terribly interesting, and the burnt-out wreck of a BP tanker lying semi-submerged nearby. We went ashore on the island first and climbed to the top of a large sandy hill to get a better view of our surroundings. With nothing at all to give any scale, this hill turned out to be a lot bigger than it had at first appeared and it took some 20 minutes of stiff climbing to reach the top, but coming down again only took a quarter of the time as the sand was quite loose, which enabled us to glissade down like scree runners.

Next up on the excursion schedule was the burnt-out tanker, which we boarded at main deck level. Even now, I find it difficult to comprehend how fierce the fire that had destroyed this ship must have been: the deck plates were buckled and split so badly that it was extremely difficult to even walk across it, while the amidships accommodation seemed to have just slumped down as if it had been made of partly melted chocolate. Through the rents in the deck we could see below into the greeny-blue depths of the tanks. These had long since lost any traces of oil and had become home to shoals of fish which could be seen quite clearly swimming in and out of the holes in the side. The after accommodation had suffered less from the heat than elsewhere and it was possible to actually get inside. We found what must have been the crews' mess; it looked as if they had been watching a film when the fire broke out, as the blackened metal film spools and the remains of the projector still lay on the deck where they had fallen. Nothing remained of the chairs and tables apart from the steel frames, all arranged facing the bulkhead that must have been used as the screen. It was quite a creepy experience, and I wondered how many seamen might have died within yards of where we were standing.

After we had loaded, it was yet another long drag back around Africa and then the full length of the Mediterranean to the Tripoli in Lebanon, where we were paid off and flew home. It had been a boring introduction to life at sea for Astrid, and I decided straight away that I would resign and find something better for her next trip. I don't regret my time with Mobil at all, as the company had always played fair and had done its best to ensure I was relieved on time and had not tried to get me to come back from leave early, as Esso had done. Mobil had also been extremely helpful when my mother had died and had pulled out all the stops to get me home as soon as possible. Finally, I had really enjoyed working with the Indian crews, who apart from anything else, gave me a taste for curry which I enjoy to this day.

12 PRODUCT CARRIERS

After enjoying a couple of months' leave, it was time for me to think about another job. Having heard Peter's tales of the good times he had enjoyed with Shaw Savill, I quite fancied a similar outfit – but by 1975 it was already plain that tramp steamers and general cargo were soon to become history. From what I could gather, the container ships that had replaced them didn't spend much longer in port than tankers, either, so there seemed little point in seeking employment there.

After a week or two of scouring the situations vacant columns of Lloyd's List, I spotted an advert for a company calling itself Panocean Anco, which operated a fleet of product carriers under the British flag and was looking for certificated engineers. On making some enquiries I found that it ran to some very varied and interesting locations, although its ships, too, seldom expected to remain in each port very long. The money, however, was first class, and even more than I had been getting in Mobil, so I thought I would give it a try. I duly sent in a CV and awaited results: these were not long coming and I soon found myself up in London attending another interview, at which I was offered a third engineer's job straight away. Panocean Anco liked to think of itself as a progressive company and had decided that third engineers would now be called second officers (technical), to put them on a par with the second mates who had become second officers (deck); and that is in fact what was recorded in my discharge book. Similarly, second engineers would become first officers (technical) and so on. I was pleased, though, that the chief engineer was still the chief.

The firm of Panocean Anco Ltd had evolved in an extremely complex manner through a series of takeovers and mergers. Athel Line, Tate & Lyle Shipping, Anco Tanker Services, Trident Tankers and P&O Bulk Shipping had all been part of the process at one time or other, and the company of Panocean Anco itself was established in 1976. All the existing ships were renamed, the former Athel Line vessel names prefixed by *Anco* and those from the P&O side by *Post*. My first ship was to be MV *Post Runner*, one of the smallest ships in the company at only 8,700 dwt. She was also to be my first motor ship.

Product carriers (or chemical tankers as they are sometimes called) are basically small oil tankers modified so that they are able to carry virtually any liquid cargo. All the ones I sailed on had the same general arrangement, with the cargo space sub-divided by 2 longitudinal bulkheads and 14 transverse ones, which gave a total of 39 tanks. Some of the more modern ones also had four cylindrical stainless steel tanks which stood on the main deck – two on each side of the flying bridge. Whereas an ordinary crude carrier will have plain steel tanks, all the tanks on a product carrier are treated with some protective coating to prevent any reaction with the cargoes, which were quite often corrosive. On *Post Runner* most of the tanks were either epoxy resin-coated or zinc silicate-coated, depending on what the cargo would be. The different substances we carried would be too numerous to list here, but included palm oil, coconut oil, refined lubricating oils, tallow, chemicals used for making plastics, solvents of various descriptions and strong acids and alkalis – the latter usually carried in the stainless steel deck tanks.

Cargo management was a full-time job, and the chief officer, who was responsible for it all, did not keep watches. Not only did the he have to ensure that each product was compatible with the coating of the tank into which it would be loaded but also that it would not react with any traces that might remain from the previous load. Some cargoes needed to be kept within a set temperature range, so these had to be loaded into tanks containing steam-heating coils, and required careful monitoring. Tank washing assumed a far greater importance with refined products, so detergents and other chemicals often had to be used with the wash water to ensure that the required degree of cleanliness was obtained. Just to ensure they were keeping on top of the job, it was quite a common occurrence for a shoreside chemist to visit the ship when we were in port, to check the cargo for contamination and also to ensure that the tanks were fit to receive it. I had occasionally been envious of the deck officers, who in most cases did not lead nearly such an arduous life as the engineers, but I am quite happy to admit that the chief officer on a product carrier had an extremely tough job and I wouldn't have wanted it for double their salary.

Apart from the strong acids and alkalis, most of the oils and solvents we carried were not particularly hazardous providing reasonable precautions were taken. One or two of them, however, were so toxic that positive pressure breathing apparatus had to be worn on deck whenever they were being loaded or discharged. One of the most dangerous substances we carried was acrylonitrile, a chemical used in plastics manufacture; under certain conditions it can give off hydrogen cyanide. Wearing breathing apparatus (BA) for any length of time is itself a pretty unpleasant business, and is yet another reason why I wouldn't want the chief officer's job.

Although cargo handling was not the engineers' responsibility, we could still be called upon to work on the pumps and pipework, so we too needed to be aware of the hazards. The company therefore sent all the officers – both deck and engine, on a three-day chemical tanker course in Southampton, which turned out to be most interesting and instructive. One topic that sticks in the mind concerned the use of the first aid chest carried on all the company ships, which contained a great deal more than the usual bandages, plasters and burn dressings to be found on most other vessels. There was an amazing assortment of bottles and vials, all with individual instruction cards, which provided antidotes to the many and varied toxic cargoes, in case some poor devil got poisoned. The acrylonitrile remedy consisted of two ampoules to be administered in the case of an unconscious casualty. Ampoule A was to be broken open onto a pad and held under the victim's nostrils to be inhaled. If there was no immediate response then ampoule B was applied in the same fashion. Our instructor then cheerfully informed us that if this didn't work either, then we could put all the stuff away again as the patient would by then be dead!

Following the course, I was to be flown out to Rotterdam to join the ship and was required to have served a satisfactory one-month probationary period on board before Astrid would be allowed to join me. As the ship was due to be in West Africa by then, this meant that she would be flying out there on her own – a rather worrying prospect, and one which we both thought was quite unnecessary. Rules are rules, however, and I did not want to be seen to be making too much fuss right at the start of my career in the new company, so we said a miserable farewell to one another and I set off to Gatwick.

Having signed on, I introduced myself to the chief, a cheerful bald-headed Australian with a very pronounced accent. He started off by saying that he understood that I had a second's ticket, whereupon I had to point out that mine was a steam ticket not a motor one, and that this was in fact my first motor ship. I was thinking that this might have been a bit of a disappointment to him, but this was not the case at all, for he exclaimed: 'Ah good, a steam man at last! You can go and sort out the boilers – they're in a right shit state as that last idle bastard third didn't do sweet FA!' This sounded rather an ominous start to the trip and he must have seen my glum expression because he went on to explain that I wasn't to have to worry about learning the diesel side of things as I would soon pick it up, and in any case, he was going to put me on the 8 to 12 watch with the best of the Chinese engine ratings, who could run the job while I did the boiler work. This sounded a bit better, and I trotted off to my cabin to get unpacked and make a start.

The cabin was very nicely fitted out with old-fashioned mahogany furnishings and had a decent en suite toilet and shower. One thing that struck me as soon as I walked in was that there was a pair of small BA sets hanging on hooks by the cabin door. These were there so that in the case of a serious toxic chemical spill they could be donned and would provide enough air to be able to reach the safety position. Compared with a normal BA set they were quite light, and were provided with a simple carrying strap which passed over the shoulder, rather than the full-size sets which were worn on the back with a full harness. They contained enough air for the average person to breathe for about ten minutes, which was supposed to be enough to reach the designated safe area on the after deck. Everyone joining a Panocean ship for the first time was given a demonstration of how and when to use this equipment almost as soon as they stepped aboard. There were also regular escape drills when they had to be worn, as they would have been in a genuine incident, and everyone (including the wives) had to take part – when Astrid eventually joined the ship, she swore that these drills were especially timed for when she had just stepped into the shower.

Dinner in the mess was excellent, with a menu containing both European and Chinese dishes. Unlike my experience in Mobil with Indian cuisine, I have found that Chinese food ashore never seems to be nearly as good as its shipboard equivalent, and I very seldom visit a Chinese restaurant, however good it is supposed to be.

When I first entered the engine room, the most lasting impression was of the huge 8-cylinder Gotaverken diesel engine, which seemed to fill up most of it. This engine stood at least as tall as my parents' house in Shirley, and weighed well over 600 tons. To help give you an impression of the size of this monster, when I later visited the bar for the first time I found that some of its worn-out exhaust valves had been turned into bar stools; their stems had been set into short lengths of pipe welded onto the deck, and they had been topped with circular cushions. From the control position at the bottom on the port side, two flights of steps were needed to reach the top: the first of these led to a walkway midway up the engine and provided access to the fuel pumps and camshaft, then the second reached the cylinder head level, where there was a wide platform going right around the engine. An overhead crane was positioned directly above the engine centre line, so that cylinder heads and pistons etc. could readily be lifted in and out.

Despite its huge size, it was a very simple machine to work on. As each cylinder was a separate unit and could have its fuel and water cooling shut off when required, if one of them developed a fault it could be isolated and the engine would still happily run on the remaining seven. Later on in the trip, when we started overhauling cylinder units, I found that it was possible to remove a cylinder head and piston, fit new piston rings and an exhaust valve assembly and have the whole lot boxed up again in about five hours – less time in fact than doing the same job on the average car!

This was a two-stroke engine with a single exhaust valve in the top of each cylinder head, and there were no inlet valves, this function being taken care of by ports cut into the cylinder liners, which would be uncovered by the piston after it had completed about three-quarters of a stroke down from the top. As the incoming air through these ports assists in removing, or scavenging, the exhaust gases from the previous power stroke, marine engineers call these ports scavenge ports, not inlet ports. This was a turbocharged engine with a single huge blower at one end of the giant exhaust pipe, while additional scavenge air was provided by the pistons themselves, which would compress the air beneath them in the scavenge space around each port every time they descended – this was known as under-piston supercharging. When the vessel was at sea it was fascinating to stand on the top level and watch the exhaust valves and rockers engaged in their perpetual dance – it gave the giant engine a lifelike quality compared with most other internal combustion engines, where there is usually little or nothing at all to be seen moving.

At the control position there was very little in the way of instrumentation compared to a steamship – just a few pressure and temperature gauges, a tachometer and some warning lights, along with a big brass engine telegraph. There was also the usual desk for the log and movement books, a blackboard displaying notes on anything out of the ordinary, what the fuel oil centrifuge was doing and which fuel tank was in use etc. To communicate with the bridge there was a very old-fashioned hand-cranked telephone which made a squealing noise every time the bridge rang down.

Starting and controlling the huge engine was done by a single large wheel mounted on a stand next to the desk and telegraph. It had a pointer that lined up with an engraved scale behind – when stopped, the pointer was at the 12 o'clock position. To start the engine ahead, the wheel was turned anticlockwise to about 11 o'clock, which would put starting air on the engine. A couple of seconds of air, accompanied by some hissing and rumbling, would get the engine moving, and then the wheel could be turned a bit more, which would shut off the air, open the fuel and cause the engine to run. Turning the wheel further anticlockwise increased the speed to whatever was required – dead slow ahead was only 25 rpm, while full speed at sea was 120, with the wheel at approx 8 o'clock. To reverse the engine the wheel was first moved back to the stop position and then turned in similar fashion clockwise – this time there would be a slight delay and some mechanical clonking, as the whole camshaft had to be moved horizontally to the astern position first (there were two sets of cams on the same shaft – one each for ahead and astern). I wish I had been able to see the camshaft opened up, to find out how the various cam followers accommodated the cams moving beneath them without getting mangled.

The remaining spaces around the main engine contained the usual collection of pumps, heaters, centrifuges for the fuel and lubricating oil, and a relatively small condenser for the deck steam system, together with three 4-cylinder Moss diesel generators. In a separate compartment above and behind the main engine was the boiler room, which housed three vertical fire-tube boilers working at around 150 psi. This was a dark and gloomy space and looked seriously in need of some TLC.

This, then, was to be home for the next five or six months. Learning all the pipework and systems didn't look terribly difficult, but I could see that there was going to be plenty of work. Going down below on my first watch I was greeted by the Chinese rating I had been paired up with, who had brought his pet African Grey parrot down with him. The bird was perched on the blackboard above the desk where it spent the entire watch, apparently immune to the heat and noise of the engine room. The Chinese are not normally renowned for their love of animals, but this parrot went everywhere with its owner and was generally considered to be an honorary member of the engine room staff, as it spent as much time down below as any of us. It was an excellent mimic and could do a superb imitation of the sound made by the engine room telephone – for the first few watches I went to answer that phone on a number of occasions, only to find I was being summoned by the parrot.

It did not take me long to realise that my rating (who was actually the equivalent of a petty officer) really knew his way around the job and could be trusted to carry out many tasks that I would normally have done myself, such as transferring fuel, starting and stopping the generators and operating the centrifuges. He was an excellent fitter and machinist, and also one of the best welders I have ever seen at sea. With his skills he could have been chief engineer on a Chinese ship, but apparently he preferred the freedom he had working for the capitalist West as an engine room rating rather than being in a more senior position under the communist regime.

My first priority, seeing as I had been given the job of sorting the boilers out, was to take some boiler samples and see what I was up against. My first sight of the boiler test cabinet was somewhat disconcerting – although it lacked cobwebs it was quite obvious that nobody had been near it for months. Everything inside was a mess of spilt chemicals, broken glass and dirty sample bottles, so I had to spend an hour or so tidying up and obtaining fresh reagents from the stores. This done, the next job was to get the actual samples, and once again this proved to be quite a challenge, as all the sample cocks on the boilers were seized solid with scale and needed a lot of wiggling and copious applications of release oil to get them moving again. Having finally obtained my samples, I repaired to the test cupboard and started the usual round of tests as described previously. There was absolutely no alkalinity present in the sample, and when I proceeded to the vital chloride test, I titrated two full burettes of the silver nitrate into the sample without a hint of the required colour change to orange. I got the same result with the other two boilers as well, which meant that all three were very badly contaminated with sea water.

I was determined to find out exactly what the chloride readings were, so my next move was to get some distilled water from Sparks (who had charge of the emergency radio batteries) and dilute my samples to give a 10 per cent solution. Another round of tests produced the same results. Finally, I took my 10 per cent solution of sample and diluted it again to give 1 per cent, and this time when I did the tests, I could actually get some readings. I have mentioned before that sea water contains on average 35,000 ppm of salt – and my worst boiler had 56,000! Even the best reading, which came from the least-used boiler, came to 15,000. If you're wondering how there could be boiler water which contained more salt than the sea itself, the answer is very simple: when water boils, it is pure water that evaporates off and any contaminants stay behind, so if the feed water being pumped in is anything other than pure water, the contamination levels will simply keep on rising unless some of the boiler water is occasionally blown out and replaced with fresh.

All this contamination had to be coming from somewhere, and my guess was that it was from the steam heater for the tank cleaning water. The sea water used for tank washing was pumped through the heater at a very high pressure – more than the steam-heating pressure at any rate – so a leaking tube would allow it to get into the steam system and eventually through the drain cooler and back to the boilers. Another test on the returning condensate confirmed this.

After my watch I found the chief and gave him the unhappy news, with the suggestion that we shut down the worst-affected boiler, open it up and give it the best cleanout we could, before refilling and chemical dosing. We would also need to shut down the tank-washing heater as soon as possible and plug the leaking tubes. The chief accepted all this and mentioned as an aside that the previous third had obviously been fiddling the test results in the log book because they were nothing like mine.

As it was going to be a few more hours before the heater could be shut down for repairs, I agreed with the chief that I would open up the offending cooler during my evening watch, when we were still on the berth and really only had the generators to look after. When I came down below at 8 pm, I found my Chinese rating had a cup of tea waiting for me, after which I had the usual walk around to check out the job. Finding all was well, I collected a few spanners from the workshop, ready to open up the cooler. Having shut all the valves and drained it down I started removing the 30 or so end cover bolts, under the rather curious gaze of my rating. I had only done about

Somewhat surprised at this, I simply bowed my head in agreement and left him to it.

This taught me an important lesson about working with Chinese crew in those days: unlike their Indian equivalents, who would do anything to please, these guys did not like to be ordered to do anything if they suspected for an instant that the officer was unable or unwilling to do it himself. Under these circumstances they would suddenly forget they knew how to speak English, and generally make life very difficult. If, however, you demonstrated that you were ready to get your hands dirty too, then they were most obliging. After that, I got on splendidly with my chap and whenever there was a job to do we would either do it together or I would start it and he would finish up. This made it a pleasure to go on watch. On one occasion he was beavering away in the workshop boring out a generator bearing on the lathe (he was a much better turner than I was) and I brought him a cup of tea for a change, saying 'This my job,' which made him roar with laughter – 'You very funny man but good officer' was his reply – which left me feeling quite chuffed.

Going back to the heater problem, as soon as we had the covers off we opened up the seawater valve to see if there were any obvious leaks. There was no need for the ultraviolet lamp this time, as the water poured out of one of the holes as if the tube had been missing altogether. Having plugged this one at each end, we checked carefully but could find no other leaks, so we boxed it up again without delay – and somewhat to my surprise it gave no further trouble during the trip.

With the contamination problem cured, it was time to set about cleaning out the boilers. We drained down the least-affected boiler first, before refilling it with fresh water, liberally dosing it with treatment chemicals, and then putting it back into service. The worst boiler was also drained down in preparation for cleaning. As soon as it had cooled down we started removing the mud hole doors around the foundation ring of the boiler. This job proved to be very difficult, as the bottom foot of the water spaces consisted almost entirely of scale, which stopped us from knocking the doors in as per normal practice. Eventually after much wiggling and hammering, assisted by the odd swearword, the job was done, and we could start scraping all the rubbish out. What the boiler really needed was a thorough chemical cleanout, but we had no facilities on board to do this, nor could we hang around waiting for a shoreside firm to come aboard and do it for us. I thought, too, that in view of the all the scale we had found there was a possibility that the furnace shell had been overheated, so I removed the burner assembly and crawled inside to have a look. There were distinct signs of burning in one area and some suspicious-looking white streaks coming from a point about a foot above the foundation ring – almost certainly a crack. In my book, this meant the boiler would have to remain shut down until it could be professionally repaired and re-certificated by the Lloyd's surveyor.

When I reported the dismal news to the chief, he said perhaps we could have a go ourselves, which rather horrified me, and I said that I hoped he had a suitably coded welder on board to do it (knowing this was very unlikely). Anyway, it was his ticket on the line if anything went wrong, so I had to go along with it. My rating turned out to be the 'coded welder', and he set about the crack with an angle grinder and grooved it out to about 75 per cent of the plate thickness before welding it up run by run until he was satisfied – it was the best welding I had ever seen. I suggested to the chief that perhaps we should apply a hydraulic test before firing it up again, so that we could at least show we had taken all due care. This was agreed, and I found a test pump in a nice wooden box on a shelf in the workshop which had obviously never been used. We pumped up the boiler to one and a half times the working pressure and found one or two other leaks at various valves and fittings; but the weld remained sound. In view of all the residual scale, however, we kept this boiler as the permanent stand-by and only used it when absolutely necessary.

While these repairs had been taking place the ship had been making one or two short trips between different berths in Rotterdam, and when finished we sailed off to Billingham on Teesside for our final few loads before departing for West Africa. Our stay in the UK was very brief – in on one tide and out on the next – and there was no shore leave, although I did manage to find a phone box on the quay to check up that things at home were okay.

After sailing down the North Sea and turning right at Dover, we were on our way down Channel when I came below for my evening watch. After the usual scrutiny of the log – not much to see compared with the log on a steamship – I had the usual cuppa with my rating and went for a walk around the job, after which I settled down to my normal chore of taking boiler samples and chemical dosing as required. This took up another hour or so, after which I returned to the control stand for another tea. I had barely taken a mouthful of this when the most awful crashing and banging noise caused me to drop the cup and make a lunge for the engine control wheel, as I was sure that the giant main engine in front of me must have been the cause of it – but unfortunately, even as the engine slowed down and stopped, the noise carried on. Realising now that it had to be one of the diesel generators that was making all the racket, I ran around the engine to take a look and found the No. 2 machine in the process of self-destruction. Even as I watched, there were bits flying off it. The stop handle was at the far end and there was no way I was going to go past it to get there, so I had to go back all around the main engine to approach it from the safe end instead. Just as I got my hand on the handle, there was an extra-large bang as one of the crankshaft balance weights flew through the side of the crankcase and went spinning off across the floor plates until it was stopped by the ship's side, after which the ship blacked out and silence reigned once more.

There was no need to ring the engineers' alarm on this occasion – even out on the main deck the noise would have been plain to hear. Having been right at the source, my head was ringing (and carried on doing so for several days). Before long all the engineers were down including the chief, who came up to see if I was okay. He was a decent sort and seeing how badly I was shaken, sent me up to have a break while the second engineer ran up one of the surviving gennies and got the ship under way again – although it was some time before the parrot could be persuaded to resume its perch on the blackboard!

When I returned a little while later, the chief was still there, surveying the ruins of the engine and noting down what was probably going to be the biggest spare parts order he ever had to send off in his life. Seeing his glum expression I said, 'Never mind, Chief – there are still a few steamships around if you fancy a change from all this!' He said he had been at sea for over 25 years – mostly on motor ships – and had never seen anything on this scale happen before, while I had been on one for just a fortnight and it had happened right in front of me. Hoping he would not thereafter regard me as some sort of Jonah, I completed my watch and retired to bed with a couple of aspirin, which I hoped would stop the ringing in my ears.

It transpired that a pair of crankshaft balance weight bolts had sheared, after which the weight had flogged around inside the crankcase until it finally broke free. Anyway, the crankshaft was now a write-off, together with that particular cylinder and all its running gear. We spent the next couple of weeks stripping the engine down to the bedplate, and by the time we had finished the whole of the deck around it was completely covered in bits which all had to be securely tied up to stop them sliding around, so that it became quite an obstacle course to get at anything on that side of the engine room.

Meanwhile, the company had been trying to find us another crankshaft, together with all the other necessary spare parts, but this proved to be a very difficult undertaking as the engines were pretty well obsolete. When the new one finally arrived about two months later, we started building everything up again but it was a slow process: the old crankshaft had been reground at some time in the past, so none of the undamaged bearings fitted the new shaft – they all required boring to size and then scraping to a fit. Each time we wanted to check the fit of the main bearings, the whole crankshaft had to be lifted out and blued, before trying it in and out again to show us where the white metal needed removing. We needed about eight tries at this before we were satisfied, and by the time I left the ship some five months later the job had still not been quite completed. Fortunately the ship's electrical load was not great and one genny could handle it quite easily, except when we were on stand-by, when the second one would be run up as insurance against a blackout when manoeuvring.

On the plus side, the boilers gave no further trouble for the rest of the trip, although it took several months' worth of blowing them down, refilling with distilled water and re-dosing before I started to get some normal-looking chemical test results. At any rate, the old chief must have been well satisfied with my efforts because he gave me a nice bottle of single malt as a leaving present when he finally left the ship.

By the time we arrived in West Africa, I had completed the required probationary period with a favourable report from the chief engineer, enabling Astrid to join me. Apart from the occasional trip to Norway to visit relatives, this was the first time she had travelled anywhere abroad on her own, and darkest Africa was not exactly the ideal destination. She was going to join the ship at Douala in Cameroon, and her flight was via Lagos in Nigeria, where she had to change planes. Never having experienced a country where nothing can be done without a backhander or tip of some sort (locally called 'dash') she soon found herself in trouble at Lagos airport.

Firstly, it was a case of getting herself through arrivals and back into the departure lounge which caused a problem. With great presence of mind and the use of the little school French she had, she was able to pretend she was one of the French air hostesses (no doubt her long blonde hair and smart suit was of some help in creating this illusion) and she eventually managed to get through and board the Lagos–Douala flight. The next problem was that she could see her suitcases still sitting forlornly on a trolley by the departure gate even though the plane was nearly ready to leave. This time she managed to catch the co-pilot as he was walking through the cabin and point them out to him, after which he got off again and soon had the sullen porters loading them aboard. Fortunately, at Douala she was met by the ship's agent and conducted safely to her hotel, where he informed her she was the first white woman he had seen make it through on her own for a couple of years.

The hotel turned out to be quite civilised, although she got an electric shock the first time she touched one of the taps in her room and it had the odd lizard running up and down the walls. She had a couple of days there in which to explore the town before the ship was due to arrive, and she had another stroke of luck when she met an expatriate Brit in the same hotel, who showed her the ropes and which places were safe to visit – we still have the beautifully carved wooden face that she bought in the market there. When she finally came aboard the ship for what was an emotional reunion, I was horrified when she told me what she had been through, and I resolved that in future we would travel together or not at all.

From Douala, our next destination was Matadi, the main port of the Congo, lying about 95 miles upstream on the river of the same name. The Congo is a huge and immensely powerful river, and our little ship made slow progress going upstream against the current, taking nearly all day to get from the estuary to the somewhat ramshackle jetty in the port. After we had tied up, it was not long before the port officials came aboard to complete the paperwork. The usual form on these occasions would be for the harbour master or his representative to be taken to the captain's cabin, where the various forms would be signed and a bottle of Scotch and maybe a carton of 200 Senior Service handed over as token of friendship. In Matadi, however, the harbour master (if that is who he actually was) marched up the gangway wearing a uniform dripping with gold braid and displaying more medals than an Italian Field Marshal. Bringing up the rear were four unsmiling soldiers, all armed to the teeth. When they left about half an hour later they took with them as much of our duty-free store as they could carry – at least eight cases of spirits and around 10,000 cigarettes would have been my guess. However, when confronted by four assault rifles and a few machetes one could hardly expect the captain to argue.

This was of course the Democratic Republic of Congo (formerly the Belgian Congo), which has had a particularly unhappy history even by African standards, so perhaps this episode was unsurprising. Needless to say, no shore leave was permitted, even if anyone had wanted it, and we were all relieved when the cargo work had finished and we were on our way back downstream – at least twice as fast as before, now that the current was with us.

From Matadi we went to Abidjan, the capital of the Ivory Coast and a former French colony. This was a much more civilised town and we did spend a few hours ashore here doing a bit of sightseeing, buying a few souvenirs and admiring some of the older French colonial-style buildings. From Abidjan we were sent back to Europe and fetched up at Coryton on the Thames Estuary. We had a couple of days there, so

Astrid was able to get home and see her parents which was an unexpected bonus for her. From Coryton we went across to Rotterdam to finish loading, and then it was back out for another round trip to West Africa.

On this voyage I know we visited Douala once again, because I remember Astrid taking me to the hotel where she had stayed before joining the ship. I think there may also have been another port we visited, but some of Astrid's letters from this round trip never got back home and I can't remember what it was. What we both remember (and heartily wish it had not been the case) was that our final port of call before returning to Europe was Lagos, which Astrid had experienced just a month or so before.

Nigeria is another country that has seen more than its fair share of human misery, and upon arrival there we had to remain at anchor outside the harbour for a week until our berth became available, in company with hundreds of other ships. These ships were the most motley collection of vessels I have ever see – including many old-style old tramp steamers, and judging by their tall, slim smokestacks, quite a few of those would even have been pre-war coal burners.

How had this grand fleet accumulated? From what I can gather, the story had started in the early 1970s when the country was rapidly becoming rich from the discovery of oil and the government wanted to expand its infrastructure as quickly as possible with a huge building programme. To facilitate this the port, which could at the time only handle one large ship at a time, would have to be massively expanded to deal with the expected increase in trade. The government commissioned a survey which said that about 5 million tons of cement would be needed to upgrade the port facilities with new jetties and access roads etc. Unfortunately several different government ministers took it upon themselves to procure the cement, with the result that in the end orders for some 22 million tons had been placed, all brought in by ship.

As only one ship at a time could unload on the existing general cargo berth, many of the ships had to wait months for their turn (years in some cases) and in many of them the cement had by then already started to solidify, so was useless. This was of little concern to the owners of the ships concerned, as they probably thought they would earn more money in demurrage (compensation) than they would have got for simply delivering the cargo, and they had chartered any old rust bucket they could find that could actually get to Lagos under her own steam. Once the ship had arrived and dropped anchor, most of her crew would have been paid off and sent home, leaving just a skeleton crew on board to keep the ship manned and ready should her turn eventually come up to enter harbour. A more soul-destroying experience for any seafarer I can't imagine.

The amount of money the Nigerian government paid out to keep the ships waiting has never been disclosed, but it would have been an enormous sum. In the end the government simply absolved itself from all responsibility for the waiting vessels, many of which eventually had to be sunk along the coast outside the port, as the cost of chipping out the solidified cement would have exceeded their value. They remain there to this day.

We were lucky in that the oil terminal was a separate facility from the general cargo berth, and there were only a couple of ships in front of us. When it was finally our turn we started discharging our cargo, which consisted of several different products. Everything we discharged from the ship used a single pipeline, so you may be wondering how the different grades of oil and other products were prevented from contaminating each other. The means of doing this was surprisingly simple and did not involve much actual cleaning of the line – although I believe that this was occasionally done, if for instance a food grade vegetable oil had to follow on from some noxious chemical. At the start and finish of the pipeline was a short half-section of pipe that could be removed so that a free piston, called a pig, could be inserted into the line. The pig was a close fit inside the pipe and had rubber lip seals. When the removable section had been replaced, compressed air was injected behind the pig, driving it all the way down to the other end of the pipeline and pushing all the residues from the previous load in front of it. Once it had arrived safely at the far end it could be removed and the pipeline was ready for the next load.

We discharged our first grade of oil without difficulty and the pig was sent down the pipeline – but it got stuck somewhere along the way. In these circumstances the obvious solution would have been to apply air from the other end and send it back again, where it could be checked to make sure the seals were okay. On this occasion, however, some bright spark (not any of our crew, I hasten to add) decided that this was going to take too long so he simply sent another pig down the line to push out the first one. This too got stuck. Finally, a third was sent in from the other end – and surprise, surprise, this also got stuck. The only option now was to start digging up the pipe at various intervals (it went underground for most of the way), breaking out a section at a time and trying to find the missing pigs. As the pipeline was about half a mile long this was obviously going to take some time, so our ship was sent out to the anchorage again – and we had to wait for what seemed like an eternity before we could start again. In total, we spent six weeks at anchor outside Lagos.

Our time out at the anchorage was boring in the extreme, and after a week we had caught up on all the outstanding maintenance jobs apart from the generator rebuild, which was still progressing, albeit slowly. Another thing that was keeping us all on tenterhooks was that it was now early December and quite a few of us were expecting to be relieved the next time we got back to Europe, in time to spend Christmas at home. Yet as the days kept going by with no sign that we were going in to unload, our hopes slowly evaporated until it became obvious that we would be spending Christmas on board instead.

The ship's agent came out by launch every few days to bring the mail and any fresh produce we required, and also to change the Walport box so we would have some new films to watch. He also arranged a few trips ashore for those wishing to take in the sights such as they were, and Astrid and I took the opportunity to go on one of these. I can't remember anything much of interest and the whole town was appallingly squalid – even the hotel where we were dropped off and which was supposed to be one of the smartest in town, was very down at heel, and we did not feel inclined to put our stomachs at risk by sampling anything off the menu. I certainly remember the taxi ride back down to the jetty, however, where the driver started off before I was even inside, which meant I had to make an undignified leap for the back seat as he was driving away. I am pretty sure that this was a spur-of-the-moment abduction attempt; at any rate the driver looked very displeased that I had managed to get on too, and Lord knows what might have happened to Astrid if the attempt had succeeded. After this incident we were happy to stay on board for the remaining time in Nigeria. Eventually the pipeline was unblocked and we could tie up and finish our cargo work, which only took another couple of days, following which we could at last put to sea again. I don't think anyone on board was sorry to see the Nigerian coast disappearing behind us.

Eventually we got back to Europe and we were paid off in Le Havre. *Post Runner* had been a happy little ship and one that we were both quite sorry to leave, while both of us had visited places that before then were just names we might have seen in an atlas. When you read about the grinding poverty that so many people have to endure it is one thing and you may be moved to donate a few quid to whatever charity has brought it to your attention – but to actually see it at close hand is another thing entirely and really brings home the huge advantages we enjoy in the affluent west.

Following this trip, our next ship was MV *Post Charger*, which was one of the newer vessels in the fleet and had been purpose-built as a product carrier, whereas *Post Runner* had, I believe, been a conversion job and carried no deck tanks. We flew out to Rotterdam to join *Post Charger* and then had a long taxi ride from the airport to find the ship, which we eventually discovered at a remote dock in the middle of a vast industrial complex. Once on board, we soon forgot the dismal surroundings, as we had been allocated a really nice forward-facing cabin with a view out over the deck on the port side. As it happened, when we crossed the Atlantic to New York following our departure from Europe, we experienced relentless heavy weather, and the view consisted mostly of green water thundering down the deck and submerging our porthole.

Post Charger was a great deal different from all my previous vessels, in that there was no watch keeping to be done. UMS (unmanned machinery spaces) was a concept that was rapidly gathering pace in the shipping industry, as it meant that the engineer officers no longer had to stand watches and so were available for maintenance. This in turn meant that fewer engineer officers had to be employed; *Post Charger* had just five, plus a lecky. How the system worked was that all the temperatures, pressures, tank levels, bilge levels, electrical systems, evaporator dump valves etc were fitted with alarms that would sound off in the engine control room and had to be answered there. During the day, there would always be someone working in the engine room to answer the klaxon and take the appropriate action, but at night the alarm would be switched through to a buzzer in the duty engineer's cabin to summon him below – and he had just three minutes to get there or else the general engineers' alarm would then be triggered, much to everyone else's annoyance. On *Post Charger* there were three of us taking it in turn to be the duty engineer, so I could guarantee two consecutive nights of undisturbed sleep – which was certainly more than I ever got on *Esso Durham*.

One other essential feature of a UMS ship is that it must have bridge control of the main engine: once the engineers had started up and checked everything was in order, we handed control over to the bridge and simply sat back and watched them drive the ship. This had the advantage for us that if the bridge managed to clout the dock when berthing we could not be blamed for a slow response to the telegraph! There was actually a conventional telegraph at the control console, so that if the bridge system failed we could still manoeuvre the ship from down below; it was tested occasionally but I never saw it used in earnest.

Personally, I never thought the system was particularly safe, as there were an awful lot of things that did not have alarms on them – motor bearings running hot for instance. On one occasion I was woken in the middle of the night and as I descended the ladder to the first platform I was horrified to find that the middle levels of the engine room were awash with water slopping over the sills around the deck edges and cascading down over the machinery below. It turned out that a rubber hose connection to an evaporator had burst, and it was only when enough water had gravitated down to the bottom levels and set off one of the bilge alarms that the buzzer went off in my cabin. On this occasion I sounded off the general engineers' alarm anyway, to get some help with the mopping-up operation. We were lucky that the main engine lubricating oil pumps had not failed, as they had both been liberally showered with water. Needless to say, with a man on watch the leak would have been spotted very much earlier and would not have caused such a near-disaster.

Another time when the buzzer went off I had just got myself soaped up in the shower, and by the time I was partially dried off and had scrambled into my boiler suit, a couple of precious minutes had already gone by, so I was in even more of a hurry than usual to get below. Upon entering the engine room there was one quite long ladder leading down to the control room level which I attempted to slide down in my normal fashion – on this occasion though, my hands were still slightly wet and slippery, so I arrived at the bottom going much too fast and twisted my ankle. I hobbled inside and pressed the Cancel button just in time, although it had sounded off for nothing of any great importance, which left me cursing whoever had designed the system.

The alarm system, although quite comprehensive for its day, was not very sophisticated: if for instance the ship was rolling and the level in a tank was anywhere near the top or bottom, the liquid slopping from side to side might occasionally trip the float switches and cause the high- or low-level alarm to sound off. It would have been so much better if say a ten-second delay could have been built in, which would have eliminated that. When it was my turn to be duty engineer I would make a final visit to the engine room late in the evening and would try to ensure that all tank levels were such that they would last until the morning without tripping an alarm – sometimes successfully and sometimes not. If the weather was at all rough, I would frequently spend the night dozing in a chair in the engine control room anyway, rather than be lying in bed awake, waiting for that extra-big wave which would set an alarm off and send me hurrying down below.

Once we had a spell of 110V earth leakage alarms that always seemed to occur at some time between 10 and 11 pm. When we got down to the control room we would find that the fault had already rectified itself and the alarm simply needed resetting. This made it impossible to trace. One day I went down to the pantry around this time to make a cup of tea and was just in time to see one of the wives sticking a knife into the toaster to get her oversized slice out of it (another one who couldn't use a bread knife!). I thought nothing of it until about 30 seconds later, when the duty engineer came hurrying along the passage, doing up his boiler suit as he went. Five minutes later he was back up again and when I asked him what the fault had been he said it was 'Another one of those ruddy earth alarms.' The penny immediately dropped and we posted a notice by the toaster instructing users not to stick anything metallic into the toaster without switching off at the socket first. We had no more earth alarms after that.

The main engine on *Post Charger* was a 6-cylinder Sulzer, not as long as the Gotaverken on *Post Runner*, but of similar height, although at 12,000 bhp it was quite a bit more powerful. The extra power was obtained by having two large turbochargers instead of one, which unsurprisingly produced considerably more boost pressure, it being the nature of any internal combustion engine that the more air you can get into the cylinders at each stroke the more fuel you can burn, and consequently the more power will be developed.

There are limits to how far this process can be taken – piston and exhaust valve temperatures, along with mechanical strength, being among the main factors. Even so, it can go a very long way: the world's most powerful diesel engine at the time of writing is the Sulzer RTA96C (the 96 is the cylinder bore in centimetres!), a leviathan which in the 14-cylinder version produces around 108,000 hp and weighs 2,300 tons – the crankshaft alone weighs 300 tons. What makes it even more amazing is that the firm of Sulzer originated from, and is still based in, Switzerland – a landlocked country which has never had any ocean-going merchant ships.

When we arrived in New York – as usual at a jetty many miles away from the city centre, the wives on board were given a ride into town by the ship's agent so they could indulge in some sightseeing and shopping, while the engineers busied themselves down below with a main engine cylinder overhaul ('doing a unit' was what we usually called it) – a task that would take about four or five hours if all went well.

This Sulzer engine had no exhaust valves but instead had two sets of ports cut into the cylinder liners, which would be uncovered before the pistons reached the bottom of their strokes. The exhaust ports would open first, followed shortly afterwards by the scavenge (inlet) ports – in almost exactly the same fashion as a two-stroke motorcycle engine. This of course made it very easy to remove the cylinder heads, as there were no rocker arms, valves or pushrods to deal with. With the cylinder head off and the crosshead nut at the end of the piston rod removed, a lifting eye would be screwed into the piston, the crane hitched up and the piston lifted out complete with the piston rod. The actual connecting rod was left in place, having a fork end attached to the crosshead at the top, and by the bottom end to the crankshaft – this was just as well, considering that it must have weighed about 2 tons and could only be removed through the crankcase door on the side of the engine by a complex system of chain blocks and wire rope slings.

Once the piston was out, the old piston rings were removed and discarded, the grooves were cleaned out, and new rings were fitted. Meanwhile, whoever had drawn the short straw would have to climb down into the cylinder and clean the carbon from the ports – a filthy job that we all detested. With this done, measurements were then taken of the cylinder bore at different places and compared with the previous readings to see how much wear had taken place. Following this, it was usual when overhauling a cylinder unit to change the fuel injector as well, and possibly the starting air valve too, for overhauled spares. Finally, the whole lot was boxed up once again.

One thing that made the job a lot easier than on earlier engines (including the Gotaverken on *Post Runner*) was that none of the nuts on the cylinder heads, cross heads, bottom end bearings or main bearings required spanners to undo them, so there was no more holding a flogging spanner on a particular nut with your foot and hitting it with a sledgehammer. We particularly appreciated this advantage when working inside the crankcase and having to balance on slippery crank webs and the like. This was achieved by all the major bolts and studs being extended well beyond the nuts, so that circular hydraulic jacks could be screwed onto them. The jacks were then pumped up, using a hand-operated hydraulic pump, until the calculated pressure had been reached, stretching the stud or bolt enough to enable the nuts to be simply unscrewed by hand, using a small tommy bar inserted through slots in the base of the jacks into holes in the sides of the nuts. Steel is an elastic material, so provided the set pressures were not exceeded, the bolts would return to their original length when the pressure was released. Doing the nuts up again was the same procedure in reverse: screw the nut on hand-tight, then fit the jack and pump it up to the set pressure which would enable the nut to be done up some more, and finally release the pressure and remove the jack. It really was that simple, and saved a great amount of time, not to mention the occasional squashed foot following a mis-hit with the sledgehammer. For the cylinder heads, whose eight nuts all needed to be done up evenly, eight jacks were provided, all connected to the same pump, so that each stud would have exactly the same pressure applied.

These very large two-stroke engines have very large oil sumps to go with them, which often contain many tons of oil; this of course makes it extremely expensive to do an oil change. To get around the problem, use is made of the fact that the piston rods are connected at their lower ends to the cross heads, which in turn run in vertical guides and will travel in a straight line, so they can be sealed off from the crankcase by a gland. This means that any of the combustion products passing the piston rings will be prevented from entering the crankcase and the oil therein will stay clean: under normal circumstances the oil is never changed at all. It is, however, centrifuged continuously, which will remove any water that might have leaked from the piston cooling system, together with any dirt that may have got in there – the engineers' boots from when they have been inside doing inspections or maintenance probably being the main source. Running on clean oil all the time means that the crankshaft and crosshead bearings should remain in pristine condition, and in fact they usually only ever get opened up for the Lloyd's five-yearly inspections.

You may well now be wondering, if all the lubricating oil remains in the crankcase how are the pistons and cylinders lubricated? The answer is that each cylinder has a number of small oil connections (called quills) roughly level with where the piston rings will be at the top of their stroke, through which very small amounts of special cylinder lubrication oil can be injected by metering pumps. This oil (which is quite thick, and when new is the colour of olive oil) will eventually percolate down into the scavenge spaces between the bottom of the cylinder liners and the crankcase top. From there it passes through reed valves into the scavenge (air inlet) trunking, whence most of the residues drain down into a collecting tank. By this time it has become a filthy black sludge and cannot be re-used, so when the ship reaches a port with suitable oil reclamation facilities it is pumped ashore. A great deal of time and effort has been put into developing efficient cylinder lubrication systems for marine diesels in the quest for a method that will give the lowest wear rate for the cylinders and rings while using the least amount of oil, and on *Post Charger* we used only a few gallons a day.

The keen-eyed will have noticed I said that *most* of the residues drain down into the collecting tank: in fact the used oil was so thick that quite a lot of it was left behind, liberally coating the insides of the scavenge spaces and trunking. To obtain a bit more boost pressure, this engine also had under-piston supercharging, where the scavenge spaces beneath the pistons were fitted with reed valves, as noted above, which would open as the piston descended, allowing the displaced air to enter the scavenge trunking, and would close again as the piston went back up again. Periodically these valves would need changing as part of the planned maintenance schedule – I think it would have otherwise been impossible to determine when one had fallen to bits. To gain access for that job meant climbing up into the scavenge trunking from below through a small trapdoor. To anyone whose only experience of a big diesel is the sort you would find in an HGV, it might seem as if I am shooting a line when I talk about getting inside what would equate to the air inlet manifold of a truck – but it's true, and this wasn't even a particularly big engine by the standards of the time.

This was another filthy job where we all hoped that the second would pick someone else to do it. To be fair, he did have a go himself once, so we couldn't complain. When it was my turn, I stood on a pair of steps under the trapdoor, which was hinged on one side, and started removing the nuts that secured it. Once I had got them off, the door was still reluctant to open so I jammed a screwdriver into the joint to help unstick it. The door immediately swung down, accompanied by a gallon or so of the aforementioned filthy black sludge, which covered me literally from head to foot. A couple of my companions (who I suspect may have known what was about to happen) had been watching this performance and were now falling about helplessly – I really did look like something that had just crawled out of a treacle mine! They were a good bunch of chaps on *Charger*, though, and by the time I had gone up and cleaned the worst of the muck off myself – which took the best part of an hour – I found upon going down below again that somebody else had done the job for me.

After leaving New York we headed south to the Gulf of Mexico. The first port of call was somewhere near Houston and on this occasion the agent kindly provided a minibus to take all the off-duty officers and the wives out for an evening meal. The venue was modelled on what the Americans imagined a Tudor banqueting hall might have looked like, complete with lots of fake wooden beams, fake candles to light the place and serving wenches in period costume. The menu, however, made no attempt to be authentic, and consisted almost entirely of different sorts of steaks, varying in size from very large to absolutely enormous. To provide some vegetable matter on the plate there was also a free salad bar with a vast range of sauces, dips, nuts and olives as well as the usual lettuce and tomatoes. I rather fancied a baked potato with my steak and ordered one up: this was a mistake, as when it came it was the biggest potato I had ever seen and was served with at least half a pound of butter and whipped cream – even when shared out between four of us it was too much. However phoney the place may have been, the friendliness and general good humour of the staff made up for it and we all came away feeling it had been a good night out.

From Houston our next port of call was New Orleans, where we spent several days, giving us all the chance to make excursions ashore. We did all the usual sightseeing stuff, starting with the French quarter and Bourbon Street, where we were somewhat taken aback to see a scantily clad girl on a trapeze swinging in and out over the pavement from the inside of one of the many bars. We also took in a club playing traditional jazz and had a nice Cajun meal. The following afternoon, having done the tourist bit and fancying something a little different, we decided to simply go for a walk up the river bank on the top of the levee and see where it would lead us.

Charger's berth was upstream of the main town on the west bank, and after walking about a mile we were into some fairly open country. We came to a side road leading inland and went down that for a while until we came to a railway line. There was a train approaching and although we could easily have crossed in front of it (there were no barriers) we stopped to watch it go past. This turned out to be a bad move as the train was of great length, with two diesel locomotives at each end and one in the middle as well. It was travelling at not much more than walking pace, either, and took over 15 minutes to go past. While we were waiting, a police car came up and stopped beside us. One of the officers wound down his window and asked who we were and what we were doing. Upon being told, he simply said: 'Okay, but take care – white folks don't usually go down there' – indicating the small collection of buildings that lay ahead. This sounded rather ominous and we did think about turning back, but as everyone else we had met in New Orleans had been friendly we decided to carry on anyway.

Apart from a few dozen houses, what we could see of the place consisted of little more than a tiny chapel with a tin roof, a filling station and a drug store. The latter in America are nothing like Boots the Chemists in the UK, but are really more like small general stores with a pharmacy included. This one also served as the town meeting place and had an eating area where you could get a coffee and cake as well. We sat down and ordered some refreshments, and before long were in conversation with some of the locals, who couldn't have been more friendly. Perhaps it was our English accents, or else they were simply not used to people saying please and thank you and generally treating them as human beings – in 1976 America still had a long way to go with racial equality. When we had finished our drinks and cake we were even offered a ride back to the ship, but we decided to walk anyway.

Just before the ship left New Orleans we were treated to the sight of the stern wheeler *Natchez* gliding gracefully downstream – a sight that would have been familiar to Mark Twain. This particular *Natchez* (there had been six or seven earlier versions) was a replica that at the time was barely a year old. I was pleased to discover, though, that she was not just a lookalike driven by a diesel engine lurking out of sight somewhere down below, but had genuine compound steam engines that dated back to 1925 and had originally been used on another river boat, the *Clairton*.

After our enjoyable stay in New Orleans we headed south for the Panama Canal, which enthralled Astrid and gave her plenty to write home about. Although I made a total of ten canal transits during my career, it never failed to impress me either. Having cleared the canal and collected the mail by launch at Balboa, we passed under the Bridge of the Americas, turned right into the Pacific, and headed north for San Diego in California. Once again we were able to spend some time ashore here, although I can't remember exactly what we did – I think it may just have been a stroll around the harbour and a visit to a cafe or bar.

Another advantage of a UMS ship is that without the need for any watchkeepers we could all enjoy longer spells ashore, provided all the maintenance and repairs were up to date. *Charger* was a very reliable ship anyway, so when we hit port there was seldom much of a backlog of work to do – during the last trip we had overhauled two main engine cylinder units as described previously, and a few cylinder head changes on the generator engines, but those were about the only big jobs. The captain and chief on *Charger* took some trouble to ensure that everyone in their respective departments got a run ashore if they wanted it, and frequently arranged with the agent for taxis and minibuses to take us. This made it a very enjoyable trip, and I think that on *Charger* I probably had more runs ashore than on all my previous ships put together. The chief officer was the only man aboard who got less than his fair share, as he was frequently involved with complex cargo work which he did not care to delegate to the other deck officers – as mentioned earlier, some of the stuff we carried on *Charger* was very nasty indeed, and he probably felt he was the man best placed to deal with it. On our side, the duty engineer and one other (which could be the lecky) were the only officers who remained aboard in port. That would have been unheard of on a turbine ship.

From San Diego we continued up the west coast until we reached San Francisco, passing under the famous Golden Gate to enter the harbour. At the time of its opening in 1937, this was the longest suspension bridge in the world, with a span of 4,200 feet – a record it held until 1964 when beaten by the Verrazzano-Narrows bridge in New York State that came in a shade longer, at 4,260 feet. Our own Humber Bridge then took the record in 1981, with a substantial jump to 4,626 feet. As the 21st century approached there was a massive surge in bridge building and the Golden Gate has now been relegated to no. 17, but it is still a very impressive sight and must rank with the Sydney Harbour Bridge as one of the most dramatic entrances to any port in the world.

We struck lucky again in San Francisco and enjoyed a couple of good excursions ashore, visiting amongst other things, the Fisherman's Wharf area where there was plenty to see and where we had a very pleasant meal in one of the many restaurants. From there we took a ride on one of the historic cable cars, which are not self-propelled like trams but are drawn along by steel hawsers running in grooves in the road between the tracks. We also took a stroll to see Lombard Street with its eight hairpin bends, which is supposed to make it the most crooked city street in the world, and has been featured in many films including *Vertigo*, *Bullitt* and *The Love Bug*. We were sorry to leave San Francisco, but all good things must come to an end and we were indeed supposed to be working for our living and not just enjoying a cruise. Before passing under the Golden Gate again and heading out into the Pacific, we passed quite close to Alcatraz Island, which by then was no longer a prison but still a very a creepy-looking place.

Now we were bound for Japan, with Yokohama as the first stop, and the Great Circle route we were taking swung way up to the north, almost reaching the Aleutian Island chain at one point. The weather most of the way across the north Pacific was vile, with green seas sweeping the deck and thumping against the front of the accommodation block, our porthole once again being frequently submerged. Astrid fortunately, was turning out to be a very good sailor and simply carried on with her correspondence course when confined to the cabin by the weather – perhaps she had inherited some Viking blood from her Norwegian mother.

In the engine room at sea we spent most of our time working through the planned maintenance schedule: the engineers' office had a great big spreadsheet pinned up on one bulkhead with 52 columns, for the weeks of the year, and a row for every piece of equipment. The cells thus formed were colour-coded for whichever particular schedule was supposed to be carried out that week, while the schedules themselves were detailed in a thick manual. Once again, I was the only engineer on board with a steam ticket, so I was naturally delegated to all the boiler work. Whoever had been doing the job before me had obviously been a lot more conscientious than his equivalent on *Post Runner*, as my first round of boiler tests showed little that needed attention, and the plant in general was reliable and in good order.

The simpler weekly or monthly checks like cleaning filters or giving a particular bearing a shot of grease had green cells on the chart: an intermediate service, which would include stuff like compressor valve overhauls and descaling evaporator heating coils, had yellow ones, while major overhauls of compressors, pumps and centrifuges etc were red – the latter we would always try and do in port, so that a failure of the stand-by machine did not have serious consequences (as it had on *Esso Edinburgh*). Things like main engine cylinder unit overhauls were, however, based on the hours run and did not appear on the chart. On all my previous ships, although we of course had routine maintenance to carry out, most of our time was spent fixing leaks, repairing stuff that had actually failed completely or else overhauling a machine that was obviously clapped out, so this was all very different. *Post Charger* was a fairly new ship, which is why the UMS system worked; I did wonder, though, whether it would still be working in say another five years' time, when things that weren't on the schedule – like instrumentation, pipework and the electrical equipment – started to develop faults and so would be taking up a lot of engineers' time.

Our first port of call in Japan was Yokohama, where we had a run ashore and explored the shopping centres and had a meal in a restaurant – there was a lot of raw fish on offer which did not much appeal to us at the time, sushi bars not yet having taken hold in the UK. Astrid also had her first experience of a Japanese loo, which left her singularly unimpressed – 'It looks like it was designed for a five-year-old,' was her comment. The Japanese there had obviously not seen many westerners, because we often found ourselves being stared and pointed at. Astrid with her long blonde hair and me with a black beard, both of which were then extremely rare in Japan, were probably what made us worthy of so much attention. Our next port of call was Sakei, where Astrid went ashore with one of the other wives while I stayed behind and worked. It was a good job that we didn't have credit cards back then, as she found some wonderful souvenir and handicraft shops in which to spend her money, and a card might have bankrupted us.

By the time we arrived in Kobe the maintenance schedule was pretty much up to date, and as there were no more cylinder units to overhaul, we once again enjoyed a whole day ashore. On this occasion the agent was offering to arrange an excursion to Kyoto, the cultural capital of Japan, as well as taxi rides to the usual bars, restaurants and clubs. In the event, we were the only takers for the excursion and were taken to Kobe station, handed our tickets and instructed exactly where to stand on the platform for our train, which was one of the famous Shinkansen – or bullet trains as we call them in the west. The train arrived about one minute before the advertised departure time, and when it stopped we found ourselves standing directly in front of the door nearest to our seats. As soon as we stepped aboard we were met by a hostess who showed us to our places and the train started to move off again, on time almost to the second – it was all so efficient that not even the Swiss, who are very proud of their train schedules, come close. Kobe station is underground, so the view from the windows remains black until you emerge from the tunnel and as the train runs on continuously welded track with none of the 'di-dah, di-dah' of the wheels crossing the joints, it means that you have no idea how fast you might be going until you reach daylight – so when we did so it came as a huge shock to find that we must already have been travelling at upwards of 150 mph and were still accelerating.

Until 1869, when the Imperial Court decamped to Tokyo, Kyoto was the capital of Japan, and as it had no industry that was worth bombing, it escaped the destruction that befell most of the other major cities of Japan. It contains many temples, palaces and formal gardens dating back to the 16th century; before that, internal wars between various samurai factions had destroyed most of the medieval city. We took in several temples, all of which had resident monks going about their daily rituals, and once we had to wait a while for a ceremony to be completed before we were allowed inside. It was the gardens, however, that we found most enjoyable, and although we had missed cherry blossom time they were still very beautiful.

We were also taken to a handicraft centre and a workshop where woodblock prints were being made by hand in the traditional manner. Most of the artists appeared to be ex-servicemen, one or two of whom had lost an arm or a leg during World War II, and some were still wearing remnants of their uniforms. Making the blocks is a very time-consuming process, but once completed, prints can be taken from them almost indefinitely. I am not sure whether or not the enterprise was entirely self-supporting or whether there was some charitable input as well, but in any event we purchased a print of some cheery-looking Japanese in a fishing boat, which still adorns our living room wall as I sit here typing now. Although we called at two more ports in Japan after this, Kobe and our trip to Kyoto was the highlight.

When we finally left that country we headed south-west to Kaohsiung, at the southern end of Taiwan. All I knew about Kaohsiung at that time was that it was the ship-breaking capital of the world – and as by now my first ship, *Esso Yorkshire*, had been scrapped, a few of her remains might have still been there on the beach when we arrived. We had a couple of good runs ashore there, including a visit to Teng Ching Park with its very pleasant lake and many Chinese-style pagodas and shrines. Once again we attracted a lot of attention from the locals, some of whom even wanted to touch Astrid's hair. We also went for a pony ride in the surrounding countryside, Astrid having being very keen on riding since she was a little girl. For me, it was my first and only time on horseback. I was somewhat nervous to say the least, but the horse knew exactly where it was supposed to be going without requiring any guidance from me, so the outing turned out to be a very enjoyable one.

On another excursion we visited the town centre, which we thought was rather tatty although there were one or two newer shops and department stores which attracted Astrid's attention. Later on in the afternoon we found a decent-looking restaurant and went in for a meal. It was very hot and humid, so we opted to sit outside on a covered verandah, where we could observe the passers-by and get some not very fresh air, there being open drains running down the streets. While we were still there eating, a torrential rainstorm started, so that after just a few minutes the street was under water. Eventually the rain stopped as suddenly as it had started, but the street remained submerged, so, ever the gentleman, I rolled my trousers up and gave Astrid a piggyback ride to higher ground, which earned me a round of applause from our fellow diners.

Next on the agenda were the Philippines, where our first port of call was the capital, Manila, on the main island of Luzon. The Philippines at this time were under martial law, effectively under the direct control of the infamous Ferdinand Marcos and his wife Imelda. She too was equally reviled by most of the population because of her lavish expenditure – estimates vary, but she apparently owned well over 1,000 pairs of designer shoes at the time the regime was finally overthrown in 1986. The military were very much in evidence all around Manila, and we only went ashore once.

This trip started with a boat ride to a rather decrepit jetty on a dirty backwater, with a lot of garbage and other debris floating about, including a dead chicken – not exactly how we imagined a Pacific island might look! Upon our stepping ashore the agent had provided a bus to take us into town – one of the famous Jeepneys, which had originated as left-over wartime US army Willys jeeps with whatever bodywork the owners chose to fit. They were all very brightly painted and decorated, and had names like Lily of Laguna, Rock of Ages, Sweet Caroline and the like. Rust in Peace might have been more appropriate for some of them.

We were not very impressed with what we saw in the part of town to which we were taken; it consisted mostly of girly bars, down-at-heel shops and some scruffy eating places, none of which we very much fancied, although we did find one decent souvenir shop and purchased a set of wooden salad bowls and some small items of raffia work to give as presents when we got back home. We were scheduled to visit at least three or four more ports, on the islands of Mindanao and Leyte, but the schedule was changed and we found ourselves en route to Bangkok instead, which pleased us immensely. This trip really was turning into a most interesting, and free, world cruise.

I am not sure exactly where the ship actually docked when we arrived at Bangkok, but it was close enough to be within easy reach of the town centre by taxi (very cheap), and we had a super day out in the city. First up were some temple visits and I think we went to at least three. One had a huge stone reclining Buddha about 160 feet long and 35 feet high, covered in gold leaf, which was most impressive. At the other end of the scale we found a very modest-looking building which contained a single quite small Buddha. We were told however, that this one wasn't covered in gold leaf, but was actually made of solid gold and weighed over 5 tons! Amazingly, there was no security of any kind that we could see, although anyone with a hacksaw could have sawn off one of the hands in a few minutes and retired for life on the proceeds. There were animals of all sorts wandering around in these temples: cats, dogs, chickens and hundreds of pigeons roosting overhead, while outside there were waterways running down the streets filled with large fish and turtles – all happily living out their lives safe in the knowledge that no Buddhist will ever willingly kill any animal. In the evening we had a meal at the Thai Classical Dance Theatre Restaurant where the food was wonderful, although Astrid could not bring herself to eat one of the dishes, which was a bird's egg salad, as she said it reminded her of blackbirds' eggs back home. The dancing was also quite fascinating, with all the performers wearing the most intricate and beautiful costumes, although for the life of me I couldn't fathom what it was all supposed to be about.

From Bangkok, we steamed south down the full length of the Malay peninsula, then made a right turn past Singapore and up the Malacca Strait to Port Kelang, some 30 miles from the capital of Malaysia, Kuala Lumpur. Here we hired a taxi for the whole afternoon (it only cost us £9!) to take us into town to see the sights.

These days, KL is a modern city full of skyscrapers, shopping malls, motorways and monorails, but in 1976 it was still very much as the British and Chinese colonists must have left it, with many fine old buildings to admire. KL had been captured by the Japanese in 1942 when they rampaged down the peninsula, but during their occupation, which lasted until August 1945, they did little structural damage. The local inhabitants were not so lucky, however, and over 5,000 Chinese residents were murdered, while thousands more of the indigenous population were sent away to work as slave labour on the infamous Burma railway, where most of them died.

On the way into town our taxi driver showed us a rubber plantation with row upon row of rubber trees, each with a cup for collecting the latex attached to the trunks. The tree trunks were a lot smaller than I thought they might have been – no more than about 18 inches diameter – and the cups only held a pint or two, which made us wonder just how many million trees there must have been to supply the rubber for all the world's tyres. Further on we saw what was probably the beginnings of the modern town, where a lot of new construction work was taking place. Taylor Woodrow was one of the contractors, which seemed rather appropriate as it was the British who had been responsible for a lot of the earlier work too.

Among other things, we saw the Sultan's Palace, which had originally been the British Residency building, a museum of local artefacts, which included a couple of steam locomotives dating back to the 19th century to spark my interest, and a shopping centre, where we found a place to have a meal. Finally, we took a cable car ride up onto one of the surrounding hills, which gave a superb panoramic view of the town – although Astrid, not having much of a head for heights, kept her eyes closed for most of the time going up and down. On the way back to the ship the road ran alongside a railway line, where I was surprised and very pleased to see a steam engine actually working.

Next, *Charger* sailed for Penang, but we were only there for a very short time – in on one tide and out on the next – and as I was duty engineer we didn't go ashore. We retraced our route back down the Malacca Strait, and then across the bottom of the South China Sea to Sandakan in Sabah, the Malaysian state in the north of Borneo.

Sandakan at that time was not much of a place, having been almost completely destroyed during World War II – first by the Japanese, when they captured the island in 1942, and then a couple of years later by the Allied bombing that drove them out. What we saw consisted of just the ramshackle jetty where we tied up, a few warehouses and buildings on the shoreside, and some storage tanks for the coconut oil we had come to collect. About a half a mile away along the coast we could see a native village where most of the houses were built on stilts out over the sea, with some brightly painted fishing boats dragged up on the nearby beach. Later that day we took a stroll down to take a closer look at the place, and once again Astrid found herself the centre of attention. The people we met were very friendly, although obviously very poor, and quite a few of the houses looked as if they were about to collapse into the water. These days the town has been completely reconstructed and there have even been some new stilt houses built especially for the tourists.

From Sandakan it was less than two days' steaming back to the Philippines, where we were to load more coconut oil. Our first stop was Jiminez on the island of Mindanao, and all we could see from the ship was little more than the jetty with a pipeline down it to load the oil, so we didn't bother going ashore. From there we went a few miles across the bay to Iligan, which was a much bigger place, and also had some stilt houses. We spent several days there, which gave everyone, including our long-suffering chief officer, the chance to go ashore. The agent had been organising transport for everyone, but while most of the crew decided to go into town (Jolly Jack seldom being able to resist the attraction of somewhere different to get a drink!) Astrid and I opted to take a ride into the interior to see some of the sights.

What I remember most is that after several miles of driving through some very thick jungle on a very bumpy dirt track, we suddenly came upon the most beautiful waterfall with an idyllic pool beneath it. Apart from ourselves and a few of the local children swimming happily in the pool, we were the only people there. These days I wouldn't mind betting that it will have been 'developed' as an exclusive tourist attraction, with an artificial beach, sun loungers, bar and maybe even a KFC or McDonalds outlet just around the corner.

The ride up through the jungle was itself quite exciting and I can remember asking the driver whether there was much in the way of wildlife, like monkeys and snakes – he replied that there might have been once but they'd all been eaten. Seeing as the average Filipino at this time was desperately poor and frequently unemployed, I don't think he was kidding. The visit of our ship to collect the annual coconut oil harvest was obviously one of the highlights of the year for the local villagers, and on the afternoon before we departed they organised a beach party for the ship's company, which included a barbecue with roast sucking pig and various other delicacies. Unfortunately I was duty engineer and missed out on this, but Astrid went along and enjoyed herself immensely, although she could not bring herself to eat any of the sucking pig, preferring some small pieces of meat in batter. Later she was told they were deep-fried caterpillars!

Our last port in the Philippines was Tacloban on the island of Leyte a bit further north. Astrid's letters to her parents don't make any mention of it, so I can only guess that we didn't go ashore there. After Tacloban we were going home, supposedly via Portland, Oregon, on the west coast of the USA, so we had another long Pacific crossing in front of us. We were expecting to be paid off in Portland, having by then completed our four-month trip – but almost as soon as we had set sail the orders were changed and we found ourselves heading straight for Panama instead, which although quite a bit further was virtually due west and much closer to the equator, so we had fine weather all the way. Our reliefs were expected to join the ship in Colon, at the Atlantic end of the canal, which therefore meant we had two transits during the one voyage.

When we were finally paid off, we were taken to a very decent hotel, where we spent a night before catching the flight home the following day. We did go for a bit of a walk around the part of Panama City nearest to the hotel but were not very impressed – it all looked very sleazy and also felt a little unsafe, so we beat a retreat back to the hotel for a meal and a swim. The pool was exquisitely maintained with underwater spotlights, so it could be used at night for anyone who fancied a midnight swim – and as the weather was extremely hot and humid we did just that. Not far from the pool there were some quite big palm trees in which dozens of fruit bats could be seen roosting. They were the biggest bats I had ever seen, and they hung upside down in the foliage making creepy noises and looking like rather smashed-up black umbrellas. I don't know what they had been eating, but the ground beneath the trees was covered in reddish-purple droppings. Vampire bats are actually quite small but perhaps this lot had also acquired a taste for blood sucking!

That summer of 1976 was one of the hottest and driest in the UK since records had begun and when we flew back into Heathrow we could see that England's green and pleasant land had become almost entirely brown and yellow. We had not been home for more than two days, however, when the drought broke and we needed raincoats for the rest of our leave.

When I had joined Panocean, it had been agreed that I would be given the opportunity to obtain five months' more steam time, which was what I needed in order to be able to sit for a combined first class steam and motor ticket. The steam time was going to be achieved by seconding me over to OCL (Overseas Containers Ltd) to sail as third engineer on one of its ships, which were all steam turbine-powered. Before this took place, Panocean asked if I would do one more job for them, on a ship called *Anco Templar*, which was going to refit at Cammell Laird in Birkenhead. I was supposed to join the ship at Avonmouth and stay with her for the trip up to the Mersey and for the duration of the refit, after which I was supposed to be relieved. I was told I would only be away for 10 or 12 days, although in the end it turned out to be a full month.

Templar was about the same size as *Post Charger* but was powered by a big B&W 8-cylinder engine, and as she was not a UMS ship I found myself back on watchkeeping duties again until we were actually installed in the dock. Apart from all the usual jobs like hull painting, replacing the sacrificial anti-corrosion anodes and removing the shipside valves and main engine components that were due for survey, one of the other jobs on the list was to remove the tail shaft for attention to the bearings or seals. These days, most ships have tapered hydraulic couplings to connect the final section of the propeller shaft to the tail shaft, which means that once the coupling has been released – a job that can be done in not much more time than it takes to rig up the hydraulic pump – the tail shaft complete with propeller can then be withdrawn out into the drydock. *Templar*, though, had the old-fashioned flanged and bolted couplings, which meant that the last section of propeller shaft inside the ship had to be removed and stowed away to one side in the shaft compartment before the tail shaft itself could be withdrawn the same way – and of course the propeller had to be taken off first. To start with, chain blocks were rigged to take the weight, and then two hefty 12-bolt flanges had to be undone and the two shafts removed, before the actual work on the bearings or seals could even start. We weren't too concerned about this, however, as it was Cammell Laird's men who would be actually doing the job, and we were there simply to point the fitters at what needed doing, provide them with whatever spares they needed from the stores and generally oversee the work.

Within a few hours of entering the drydock and being settled on the blocks, our clean and tidy ship was transformed into a complete mess: welding transformers and portable compressors were craned aboard and dumped on the decks, temporary floodlights were set up all over the place, and the decks were festooned with welding cables, oxyacetylene hoses, airlines and the like. The yard had made some attempt to keep the passageways and mess rooms clean by taping brown paper over the floors – but in the engine room dirty boot marks and handprints were everywhere, and as the ship was on shore power, none of the generators were running and it soon felt cold and dead. Pretty soon it also started to smell bad, as certain of the workers couldn't be bothered to walk ashore to the toilets and were using the bilges instead, so there was not much love lost between us and them. I couldn't wait for my relief to turn up so I could go home.

Work on the tail shaft did not start straight away and when it did, the first set of men managed to remove just one bolt during their shift. The next lot managed two, and it became clear to us that they were deliberately spinning the job out. At that time, British shipbuilding was going through a very lean period and many long-established yards had already closed down, but it seemed to us that a go-slow such as this was not going to improve the chances of the yard's survival. The work was being organised into two 8-hour day shifts which finished around 10 pm, after which nothing was done until the first shift started the next day. The fourth engineer and I therefore decided to speed things up a bit, and when the late shift had finished and gone ashore, we went down and whacked all the remaining bolts out, bar one at each end to support the shaft in place.

When the first day shift turned up in the morning and saw what we had done, they brought in the shop steward, who called everybody out on strike. A meeting was convened, with the captain, chief engineer and myself on one side and the yard manager and various union officials on the other, where I was asked why I had done union work when I wasn't in the union. I immediately took full responsibility, saying (with tongue firmly in cheek) that I was sorry to have broken the rules but it seemed as if the yard fitters were having difficulties with the bolts and I was simply trying to help them out a bit. My explanation cut little ice with the union representatives and after some further discussion it was decided that if I would agree to leave the ship and not darken Cammell Laird's doorstep ever again they would resume work. This of course suited me just fine, and even the captain gave me a conspiratorial wink as I signed off and collected my discharge book. A few hours later I was on the train home. I never did find out how much longer it took to finish that tail shaft job.

13 BOX BOAT INTERLUDE

When our leave was up, orders came through for us to join the container ship SS *Flinders Bay* at Tilbury, in April 1977. Astrid's parents drove us there and were able to come aboard for a little while to see the ship. They were very impressed when they saw the cabin we had been allocated, which was spacious and beautifully constructed in light-coloured wood, with some nice pictures and fitted carpets. Its en suite bathroom had an actual bath installed in it as well as the usual shower, so Astrid immediately felt at home. We were told it was normally kept as a guest suite but as there were no guests on that trip we would be welcome to have it. One other advantage was that as it was situated on the starboard side of the accommodation block we could see out of the porthole – on a box boat, if you are in any of the forward-facing cabins all you will normally see is the back of a box!

SS *Flinders Bay* was one of a class of six ships built for OCL in 1969–70, and these were among the first purpose-built container ships in the world. A new container terminal had been built at Tilbury to handle the ships and this, together with the container cranes, the straddle carriers and the fleet of container lorries that that went with it, represented a huge investment which should have given Great Britain a head start in the new trade. However, as is the way with most labour-saving innovations, not everyone will welcome the change with open arms; in this case it was the Tilbury dockers who, fearing job losses, refused to work in the new terminal. This meant that during the first year the ships were in service all the containers they carried for the UK had to go via Rotterdam, where they were trans-shipped and sent to Harwich and Ipswich instead. This gave Rotterdam a lead in the trade that it has never lost.

The OCL ships were powered by a Stal Laval turbine set of 32,500 shp, and were capable of 24 knots. They each had a capacity of around 1,600 standard containers (TEU), about 350 of which could be refrigerated if required. TEU, incidentally, the universal term used to define the size of a container ship, stands for twenty-foot equivalent unit – (still not metricated!) Although containers can also come in lengths of 40 feet, the original 20-foot by 8-foot size is the most common (whatever the length, the width is always 8 feet). Today there are many vessels in service of over 20,000 TEU, so a 1,600 TEU ship would now be considered very small.

OCL had been founded in 1965 by a consortium of four long-established British shipping companies: P&O, British & Commonwealth, Furness Withy and the Ocean

Steamship Company, whose decision makers, having foreseen the demise of general cargo, had decided to invest in container ships instead. OCL lasted until 1987, by which time P&O had become the major shareholder and the name was changed to P&OCL. In 1996 the firm merged with Nedlloyd and became P&O Nedlloyd, which in turn was bought out by the Maersk group in 2006. It was at that point that the name P&O finally disappeared, although one may occasionally see a container lorry with a box that has not been repainted and still shows the old P&O markings. At the time of writing Maersk is the biggest container ship operator in the world.

Having signed on, I went to find the chief engineer to see what I was supposed to be doing, and was informed that I would be put on daywork and would not be required to work until the following morning, although I should go down for stand-by when the ship sailed later that evening. I would have gone down for a look around the engine room beforehand anyway, but as it was by then nearly dinner time, I went back to the cabin, put on my uniform and went down with Astrid to the mess instead. This company was very formal compared with my previous experience on tankers, and uniform was always worn at meal times, while the steward was extremely smart in a crisp white jacket with brass buttons and served the food from a silver salver. We could have been at the Ritz.

When I went down for stand-by after dinner, Steve Rooke, the senior second, took me to one side and asked me what I knew about turbo feed pumps. He seemed very pleased when I was able to tell him that I had rebuilt two on *Esso Durham*, and he told me that fixing another one was going to be my first job. It transpired that one feed pump had failed on the voyage home and that a maker's man had been sent to fix it while the ship was in Tilbury. This worthy had finished rebuilding the pump shortly before the ship was due to sail but when it was tested it had run for just a few seconds before seizing up, whereupon he had gone home and left us to it.

The ship was running a scheduled service and there was no question of delaying the sailing until we could get the pump fixed. The ship had only two main feed pumps anyway, plus a small electric boiler-filling pump, and now we were setting out on a voyage to Australia with just the one main pump. The chief engineer was a grand old chap and was on his last trip before retiring, so he was worried sick. I was prepared to start straight away, but Steve said to go and get a decent night's sleep first and then start fresh in the morning, by which time the pump would have cooled down. Despite the seriousness of the situation, he never put any pressure on me to hurry up, and said that he wanted a good job and not a quick one. Under similar circumstances I am not sure I could have kept as cool as he did.

The next morning I started stripping the pump down, which turned out to be a lot easier than the ones on *Durham*, as the maker's man had obligingly applied Copaslip to all the bolts so they came undone quite easily. By dinner time that day I had it mostly apart, but even with the turbine and pump end covers off the machine was still seized up, which did not bode well. After dinner I decided to give it another couple of hours and went down and made another examination to see if I could find what was sticking. At first, there was nothing obvious to be seen, and there was no damage to the turbine blading – which was a relief, as we did not have a spare shaft and would not have been able to get one until Cape Town at the earliest. Then I had a closer look at the thrust bearing, which was in two halves. The top half was supposed to simply lift out, after which the bottom half could be rotated 180 degrees around the shaft and would then come out the same way. In the event, the top half was extremely tight and took a lot of judicious tapping and prying before it could be removed, at which point the cause of the trouble was revealed – it had seized up against the thrust collar. More tapping back and forth was required, but eventually I got the bottom half out as well.

At this point, I began to have doubts as to what my eyes were telling me – it appeared that the maker's man had put the thrust bearing in the wrong way round. This meant that instead of the carbon face of the bearing being in contact with the thrust collar, it was the steel housing. It only took a few moments more to show that the component would indeed go in either way around. I was astounded that someone who was supposed to be an expert could have made such a simple mistake, but in fairness he had apparently worked over 24 hours on it without a proper break and must have been dog-tired. When I reported my findings to Steve, he didn't believe it either, until I showed him the damage on the bearing housing.

Even though we had now found the problem, the thrust collar itself was badly scored, and as noted before we didn't have a spare shaft. Starting work again the next day, I thought I might be able to clean up the collar by hand filing, and set to with a new 6-inch second cut file from the stores, rotating the shaft as I worked until I had gone all the way round several times, and finishing with fine emery cloth to polish it up. The steel shaft was quite hard and so it was slow going, but by lunchtime I had got it quite smooth again, although the deeper marks could still be seen clearly. I reasoned, though, that providing there were no actual sharp edges to tear up the carbon thrust faces it should be okay.

I had a quick lunch in the duty mess in my boiler suit and then went back down again to box it all up. It was gone 9 pm when I finished, and I telephoned Steve to see if he wanted to try a test run. He came down about 10 minutes later and I was surprised to see that the chief had come down with him and was also in a boiler suit – I could count on the fingers of one hand the number of times I had seen *that* before. Steve cracked open the throttle until steam could be heard blowing through the drains and, having then shut them, he quickly wound it wide open and the machine ran up to its full speed of in a matter of seconds, while the discharge pressure gauge whipped up to something in excess of 1,000 psi and sat there without a flicker. Steve watched it run for a short time before giving me a smile and a thumbs up, saying, 'Looks like we've got a pump.' You could almost see the worry draining away from the old chief's face, and he came over to shake my hand and clap me on the back. It was one of the happier moments of my career, and from that moment on I couldn't put a foot wrong on that ship.

Having a well-earned beer in the bar later that evening, I asked Steve whether the electric feed pump, which was used primarily for boiler filling when starting the plant, would have been enough to actually keep the ship under way. He replied that even with the turbo generator shut down to save steam and the diesels taking all the electrical load, he thought it unlikely that the electric feed pump would have supplied enough water to give any more than dead slow. No wonder the chief had been worried. Steve went on to tell me some more about the chief engineer, whose name was John Archdeacon, and Michael Champneys, the captain, who was also on his last trip. They had both seen a lot of action during World War II and wore medal ribbons on their No.1 uniforms at mealtimes. During all my time at sea I never met a finer pair of English gentlemen.

The following day I was able to have my first proper look around the engine room and was very impressed with the general layout. The main difference to my previous steamships was in the electrical arrangements: instead of having two main turbo alternators and a small emergency diesel (which in the case of the Esso tankers was barely able to provide enough power to start the main plant), this ship had just one very large turbo alternator and no less than five diesels. When I asked Steve about this he said that because of the very large power requirements of the 350-odd refrigerated containers, which might contain millions of pounds worth of frozen and chilled produce, it was imperative that in the case of a blackout the diesels could provide all the power required to keep the boxes cool. The temperatures of the containers were continuously monitored and logged so that any problem developing would be jumped on right away; preserving the cargo was always the number one priority. It was so important that once on SS *Remuera Bay*, one of the bigger twin-screw OCL ships, when they had a boiler contamination problem, the instructions from head office were to get the cargo back even if it meant wrecking the boilers! This surfeit of electrical power also meant that the main boiler plant could be shut down in port for maintenance work, as it was very easy to restart when required. When at sea the turbo would normally provide all the power required, except under stand-by conditions, when at least two of the diesels would be run up as well.

Another difference was that the Foster Wheeler boilers were of the later ESD3 pattern, where the burner assemblies are mounted on the roof of the furnaces and the flames are projected down into the furnace at an angle about 40 degrees to the vertical, instead of being horizontally mounted at the bottom of the furnace fronts, as in all my previous turbine ships. These boilers had electric igniters too, so that when flashing them up, there was no more messing about with matches and a torch dipped in paraffin; they could simply be switched on and off from the control platform. Steam pressure at the turbine inlet was 960 psi – the highest I ever saw.

In the engine control room there were schematic layouts of the fuel oil transfer system and the bilge and ballast system, showing all the valves that needed to be opened to do any specific task. Each valve had an open/close switch and an indicator button that popped out when the valve was opened and back in again when closed. By selecting another set of controls, the actual levels in a particular tank or bilge well could be displayed – all very advanced for its day. It was a pneumatic system using small-bore copper pipe, and behind the control panel it looked like a mass of copper-coloured spaghetti.

All the pipe runs connecting the fore part of the ship to the engine room went down what was called the duct keel. To enter this you first had to pass through a small watertight door at the bottom of the for'ard engine room bulkhead, which led into a very narrow and dimly lit tunnel at least 120 yards long which went right up to the focs'l spaces, where you would exit through another watertight door. This tunnel was situated at the very bottom of the ship and actually passed through the double bottom spaces, so beneath the chequered plate walkway and all the pipe runs, the steel deck you could see really *was* the ship's bottom and the sea below that was barely an inch away! It was just about the most claustrophobic space I ever saw, and although there was an escape shaft halfway along, with a 50-foot vertical ladder up to the main deck, I was never particularly happy down there. Steve, who had been showing me around, had come with me and I think he was eyeing me up to see whether I would have the bottle to go the full length of the tunnel. He said later that he had known one chap who had refused point blank to go into it at all.

As it happened I spent a fair amount of time in the tunnel on that trip, because the valves and actuators of the remote pumping systems always seemed to be sticking and needing overhaul, and a great many of them were situated in the duct keel. Fortunately, there was a duplicate control panel about halfway along the tunnel, which saved a lot of traipsing back and forth to the engine control room when you wanted to test them. The valves were made by a Dutch company, Cupedo, and we always referred to them as Cupedo valves rather than air-operated valves. Up until then I had never been particularly worried about working in confined spaces, but I can't say that I ever enjoyed working down that tunnel, especially when the ship was rolling – it was more like being in a submarine than an engine room, and the proximity of the sea beneath my feet was always at the back of my mind.

Flinders Bay was a happy ship and there was nearly always a convivial gathering in the bar in the evenings. As well as the usual film shows we had 'horse racing' with cardboard horses sent down a string and bets placed on the outcome. Being on daywork meant that I seldom missed out on any of this social activity, and Astrid enjoyed herself immensely. To make matters even more interesting for her, the captain arranged for her to take her steering ticket, which meant she had to spend a certain number of hours at the wheel actually steering the ship, maintaining a compass course and learning the various helm commands. The weather was very rough for quite a few of her spells on the wheel and she said at first that whenever the bows of the ship disappeared in spray when they hit a big wave, she was terrified that she would get the blame for any damage. The bridge watchkeepers, I think, enjoyed having Astrid as their guest helm, and ribbed her gently about steering a straight course – 'That wake's as straight as a donkey's hind leg' and similar comments – but at the end of the training she was awarded the certificate and is rightly proud of it.

Our first port of call was going to be Fremantle in Western Australia via the Cape of Good Hope and the Southern Ocean. The captain opted for a Great Circle route that would take us right through the Roaring Forties down to 50°S, where he said there might even be a chance of seeing icebergs. In the event there were no icebergs but the Southern Ocean still lived up to its evil reputation and we experienced continuous gales and some very big seas – and this was summertime down there! Container ships generally are a lot less stable than tankers and at times the ship was rolling 25 to 30 degrees each way, which made mealtimes quite entertaining. The mess room tables all had fiddles – wooden boards around the edges which could be hinged up to provide a barrier about an inch high to stop things sliding off – but the main ruse employed by the steward was to dampen the table cloths before laying the tables, using a small watering can he kept for the purpose. By this means the coefficient of friction between plate and table cloth is very much increased, so the main problem then was not to keep the plate from sliding away but to keep the food on the actual plates – boiled potatoes, Brussels sprouts and peas being particularly determined to shoot off across the table. Nevertheless, everyone still sat down in their uniforms for dinner, the steward still came around with his napkin over his arm, serving the food from the silver salver, and everyone carried on as if it was all perfectly normal. Astrid, who was never seasick, thought it was quite hilarious and said it reminded her of the film *Carry on up the Khyber*, when Sid James, playing the British governor of a fictional Indian province on the North-West Frontier, is sitting down to dinner with his wife and fellow officers in the residency building and they are all carrying on polite conversation, commenting on the wine and asking for the salt to be passed etc even though the building is being shelled to pieces by rebel tribesmen, with the windows being blown in, the piano smashed and lumps of plaster falling off the ceiling.

Flinders Bay was a fast ship and despite the bad weather in the Southern Ocean we arrived in Fremantle having steamed the 11,000 miles from Tilbury in just 20 days. Almost as soon as we had tied up Steve came up to me and told me there was a taxi waiting and that Astrid and I could have the rest of the day ashore – apparently this was a reward from the chief engineer for fixing the feed pump. In fact, at every port we visited apart from Melbourne, where we had a couple of cylinder heads to change on the diesels, transport was arranged to take us ashore – and even there Astrid did not miss out, as Steve's wife Marion was also on the ship and the two women went sightseeing together.

In the event, we didn't find much of interest in Fremantle and decided to take the train into Perth instead. February in Australia is of course the height of summer and it was roasting hot. The train ride took half an hour, and even with all the windows open we were pretty well cooked by the time we got off. Because of the heat the streets were almost deserted and we had the place to ourselves – as Noel Coward wrote, 'Only mad dogs and Englishmen go out in the mid-day sun.' We found a very pleasant park overlooking the inner harbour, where it was a bit cooler and we could keep in the shade under the trees, but eventually we too sought shelter from the heat and found a pleasant air-conditioned restaurant for a meal before returning to the ship.

Next on the itinerary was Sydney, and there we enjoyed a wonderful day ashore. We started off with a trip around the harbour on one of the many ferries and had a good look at the famous bridge and the opera house, the latter having only been open a few years at the time of our visit. We got off the ferry at Taronga Park and Zoo, which is set on a very large site overlooking the harbour, and apart from seeing the animals, the views are worth it in their own right. We were pleased to see that the residents all had good-sized enclosures, compared with what they would have had in most European zoos. Following this, another ferry took us to Manly Beach, which we had been told had better scenery and more things going for it (apart from the actual surfing) than the more famous Bondi Beach, on the other side of the harbour entrance. At Manly we spent a lazy afternoon strolling around and enjoying the sunshine, interspersed with the occasional stop for an ice cream or a beer.

Eventually it was time to return to the ship for dinner but when we got aboard we found that some of the other officers, including Steve and Marion, had arranged to go into town for a meal instead, and we were invited to join them. The chosen venue was the Old Spaghetti Factory, situated in one of the older parts of the city, close to the south side of the bridge in an area called The Rocks. Inside, the décor was all very Victorian, with dark wood panelling, old-fashioned brass light fittings and a big fan on the ceiling. It reminded me of the sort of saloon bar you might find in a pub at the centre of an English market town. Around the walls there were faked photographs showing the spaghetti harvest, with women picking the strands from spaghetti bushes then laying them all out to be sun-dried and other similarly bogus processes in spaghetti manufacture; it was all quite amusing and as the food was also very good it made a very pleasant evening.

The next day we finished loading and headed off to Wellington, on the north island of New Zealand. This meant a two-and-a-half-day passage across the Tasman Sea, which gave us a very rough time indeed. There was an inclinometer on the bridge which would record what the biggest roll was in either direction; we were rolling 30 degrees each way at times, and even recorded one of 37 degrees, which was very scary indeed and sent everything flying that wasn't actually tied down. The mess room tables were bolted to the deck and I think there may have also been some readily available method of securing the chairs too but all the furniture in the bar was loose and the whole lot had to be roped together in one corner to stop it sliding from one end to the other.

You might ask why a fairly big and well-found ship such as *Flinders Bay* should behave in such a fashion when a tanker in similar conditions wouldn't roll even half that amount. The answer is that on a tanker all the cargo or ballast is below the main deck, which makes for a nice low centre of gravity. On a box boat, however, the containers are not only stored in the holds but also stacked up on deck – almost up to the bridge sometimes – and in those days the chief officer, who was responsible for loading them, couldn't always arrange to have the heavier boxes loaded in the bottom of the ship with the lighter ones on top, which would have helped a lot. There were no loading computers available back then, and the calculations must all have been extremely difficult. Much as I dislike anything to do with computers, this is definitely a case where they are an essential tool for ship safety. Any modern container ship will have a program tailored for it, showing all the available slots. Whenever a container is loaded on or off, its weight (which will be known as soon as the container-handling crane picks it up), can be entered for the appropriate slot and the ship's new GM, which is the usual measure of stability (for more about this, read on) will be flashed up on the screen within seconds, together with a warning when it gets too low.

This book is not a treatise on naval architecture, but I will attempt to explain the basics of ship stability here. I am not going to go into why a ship floats in the first place – Archimedes worked that out several thousand years ago. It is why it floats

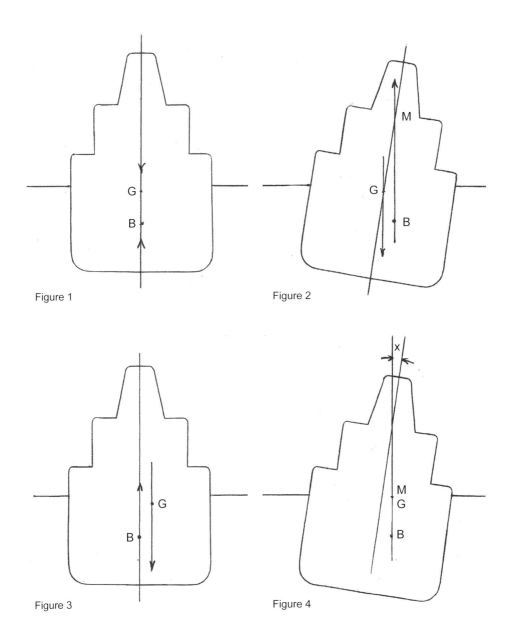

Figure 1

Figure 2

Figure 3

Figure 4

upright that concerns us. This is dependent to a large extent on where the centre of gravity of the ship is and its underwater shape. Regarding the latter, although a vessel may have a very sharp Vee-shaped bow and a rounded stern, its underwater cross-section for at least 2/3 of its length will be rectangular with rounded bottom corners – see Figure 1. At the centre of this rectangular area will be an imaginary point called the centre of buoyancy (COB or simply B on the diagrams), which will lie on a vertical centre line going right up through the middle of the ship and through which the upward force that keeps the ship afloat will be acting. A bit further up this vertical line will be the centre of gravity (CG, or G on the diagrams) which should not need explaining and is the point through which the downward force trying to sink the ship will be acting – these two forces are basically equal and opposite.

Now imagine the ship heeling over to one side as shown in Figure 2: the draught on the uphill side will be reduced and that on the other will be increased, which will make the underwater cross-section trapezoidal rather than rectangular in shape and the COB will therefore move across a bit toward the fatter end – the CG however, will remain in the same place. If you now draw a vertical line up through the new COB, it will intersect the original vertical line a little way above the centre of gravity – the point of intersection is called the metacentre (M on the diagrams).

Now we can see that the upward force is not in line with the downward force, and the resulting leverage will bring the ship upright again. The horizontal distance between the up and down forces is called the righting moment, while the distance from the metacentre to the CG is the GM. All the time the GM remains positive, the ship will try to right itself when heeled over and naturally enough, the smaller the GM the more easily the ship will roll – it is then said to be tender whereas a ship with a lot of GM is said to be stiff.

Coming to Figure 3, we have a very perilous situation, where the CG and the COB are no longer on the same perpendicular lines, perhaps because the cargo has shifted in heavy weather or a compartment on the downhill side has flooded. Either way, the GM has become negative and we can see that the buoyancy force and the weight of the ship are now trying to heel it over, rather than bring it upright – it will in fact flop over until it reaches the position shown in Figure 4. This will remain a permanent list (known in the trade as the 'angle of loll' and marked X at the top of the diagram) unless the CG can be restored to its former position – possibly by pumping fuel or water from one side to the other. We can also see that G and M are coincident, so at this position the GM is zero. If the ship were to roll just a little more, the GM would become slightly positive again, and the ship would try and return to this previous list, although it can clearly be seen that it won't have to go very far before the side of the ship goes under water, the hull will start to flood and finally of course the ship will sink.

Even now, in the age of the computer, if you want to know the GM of a ship accurately you have to perform what is called an inclining experiment. For this the ship has firstly to be loaded or ballasted to be exactly upright, after which a known weight will be shifted on deck a measured distance from one side to the other and the resultant list measured on an accurate inclinometer. A lot of calculation work ensues and finally you come up with the GM. This experiment is most often carried out on new ships so that when they go into service the chief officer, whose job it will be to calculate the loading, will have a starting point to work from. The maths required to do this is quite complex and very time-consuming, especially as in earlier times it would all have been done using slide rules or log tables (which a great many students today won't even have heard of), so just for once I will say, 'Hurrah for the computer!'

Of course, not all ships have rectangular underwater cross-sections: if you look at HMS *Victory* in Portsmouth Historic Dockyard you will see that her underwater hull section is very rounded indeed, while submarines are even worse, with almost completely circular hull forms. So where does their stability come from? The answer is that they all have very low centres of gravity – in the case of *Victory*, she has about 260 tons of pig iron overlaid with shingle, right down by the keel, while for the submarines it is all the machinery, stores and fuel that keeps the CG low enough. Older diesel-electric submarines had huge lead-acid batteries below the floor plates which did the job, although they still rolled abominably in heavy weather; one can only imagine how the World War II German U-boat crews must have felt when they had to spend every night on the surface in the North Atlantic recharging their batteries.

Cruise ships are another class of vessel which appear to defy the laws of stability, with their comparatively shallow draft and towering superstructures, sometimes ten or more decks high. Although they may be shallow-drafted, they are also quite broad-beamed by comparison with many other ships, which of course helps considerably. Additionally, they usually have aluminium superstructures to reduce their apparent top-heaviness, while in the engine room there will be at least four, and sometimes five or six socking great diesel generator sets weighing 100 or more tons each, set right down in the lowest part of the ship, which together with all the fuel and stores keep the CG well down. I bet however, that the captain's heart may still miss a beat or two when several thousand of his passengers all run across to one side to watch the Statue of Liberty or some other attraction, as they glide past!

After this little discourse, it's time to get back to *Flinders Bay*, but while I can remember in some detail what we did in Sydney, in the case of Wellington it is all very hazy. I know we had a car (courtesy of the chief yet again) and took a drive into the countryside where we visited the Remutaka Forest Park, but apart from that it is all a blank. Sailing on from Wellington, our next port of call was Lyttelton, next door to Christchurch on the South Island. Here Astrid and I were taken into town and dropped off, and we decided to hire a car and head inland toward the Southern Alps. The first car hire place we tried was clean out of cars but offered us a Toyota Hiace van at a bargain price instead, so we took it. This turned out to be a good choice because we were sitting about 2 feet higher than we would have been in a car, and had much better views of the countryside as a result.

To start with we headed inland across the Canterbury Plain, which is where a lot of New Zealand lamb comes from – and a good deal of which would be in the ship's refrigerated containers when we sailed back to Tilbury. Leaving the plain behind, we started steadily climbing as we reached the foothills of the Southern Alps. Our route took us up over Arthur's Pass and en route we saw some quite spectacular scenery, with beautiful river valleys, rocky hillsides and snow-capped mountains everywhere. For quite a bit of the way we were travelling in quite close proximity to the Christchurch–Greymouth Railway, which is reckoned to be one of the most scenic railway journeys in the world and has become a top tourist attraction. The railway was a lot less famous in 1977, however, and we didn't see any trains. Eventually we reached the summit of the pass, which is on the watershed between the east and west sides of South Island. There was a big parking area where we could stop and admire the views, which extended pretty well down to the far coastline at Greymouth.

It was past lunchtime by now and the sunshine we had enjoyed on the way up had been replaced by threatening, solid grey cloud. We reluctantly decided to beat a retreat back down the pass before it got much worse, which turned out to be a wise move, as before we had gone very far it was snowing hard and we began to worry about getting down at all. Fortunately the Hiace van turned out to be quite sure-footed and as there was hardly any traffic I was not afraid to chuck it around a bit on the descent and risk the occasional skid. Eventually the snow was left behind and turned into heavy rain instead, which was quite a relief as we had been having visions of getting stranded and missing the ship, which was due to sail early next day. Before much longer we got to a filling station and shop, where we stopped for fuel and something to eat. When we had finished we had a word with the guy who seemed to be in charge and told him how hard it had been snowing back up on the mountain, to which he replied, 'Jolly good!' When I asked whether this was because it would be good for the skiing, he said, 'No, mate – but it'll be great for my breakdown and towing business!'

After our little adventure, which had turned out to be one of the best days ashore we ever had together, the ship sailed as planned the following morning, heading for Panama and then back to Tilbury. Apart from more Cupedo valves and some diesel generator overhaul work, I can remember nothing of the journey home, which goes to show just how reliable the ship was – indeed it was the most reliable turbine ship I have ever sailed on.

Heading roughly eastwards for the whole trip meant we had been heading into the rising sun and were effectively shortening each day. On board ship we compensated for this by putting the clock forward one hour at midnight for every 15 degrees of longitude we covered. This process was called 'flogging the clocks' and as *Flinders Bay* was a fast ship we were doing it every other day. To ensure all the watchkeepers benefited equally from the short day, each watch took a 20-minute share. Naturally if we kept doing this all around the world we would have arrived back home a whole day ahead of everybody else, which of course would never do; that was why the international date line was conceived. This imaginary line roughly follows the meridian of 180 degrees longitude (there are a few hefty kinks in it around island groups, so that one island will not have a different date from its neighbours) and when you cross it from west to east, as we were doing, you actually have the same day twice. Going the other way you miss a day, so if it was for example the 23rd of the month as you were approaching the line going westwards, the following day would be the 25th. This calendar anomaly was a crucial part of the plot of Jules Verne's famous novel *Around the World in 80 Days*, where the hero, Phileas Fogg, bets a fellow member of the Reform Club that he can do the journey in that time or less. After many adventures and tribulations, Fogg arrives back in London believing that he has lost his bet by one day – but then discovers that because he crossed the date line from west to east he has arrived back 24 hours earlier than he had thought, and therefore wins instead. He should have come with us on *Flinders Bay*, which got us around the world in 73 days with no tribulations at all!

After few weeks' leave Astrid and I were back at sea again with OCL, on one of *Flinders Bay*'s sister ships, SS *Jervis Bay*, which had a very famous predecessor in

World War II. In 1940 Britain had been very short of convoy escort ships, and to try and make up the shortfall a number of merchant ships were requisitioned by the Royal Navy, one of them being *Jervis Bay*. She was already quite old (built in 1922) and had a maximum speed of just 15 knots, while her armament consisted of eight 6-inch guns that pre-dated even World War I, and a couple of 3-inch anti-aircraft guns. She had a Royal Navy captain, Fogarty Fegen, but the majority of her 255 crew were from the Merchant Navy or the Royal Naval Reserve.

So although she was not exactly in the same league as a proper destroyer she had been given a destroyer's job to do and in November 1940 she was the sole escort for the 37 ships of convoy HX84 sailing from Halifax, Nova Scotia, to Liverpool. At some time during the evening of the 5th, the German pocket battleship *Admiral Scheer* was sighted, and Fegen ordered the convoy to make smoke and scatter, while he steamed in to draw the enemy's fire. The odds against *Jervis Bay* were hopeless; she was badly damaged and on fire even before her own guns were in range, but she kept closing the enemy and firing until an 11-inch shell from *Scheer* penetrated her engine room and smashed the condenser, which immediately stopped the engine and flooded the space. Even then her gunners kept on firing, and it took *Scheer* a whole hour and a lot of ammunition to sink her – by which time the majority of the convoy had managed to escape into the gathering darkness.

Captain Fegen and 189 of his crew died during the battle, and he was posthumously awarded the Victoria Cross. The chief engineer – who, like most of the officers and crew, had stayed at his post and had also gone down with the ship got nothing, which reminded me of what my dad had said about medals 'For every man who got one, there were another ten who should have done but didn't.' Hanging on one of the bulkheads in the officers' mess of our *Jervis Bay* was a fine painting depicting the action, along with the usual portrait of the Queen. By a strange coincidence, a few years later when I was a chief engineer on the cross-Channel ferries, I sailed with a second engineer whose father had been one of the few survivors of the battle.

This second trip with OCL was very much like the first, except that this time we did it in the reverse direction, heading out to New Zealand via Panama before going on to Australia and then back home. Regarding work, once again the ship was very reliable and we had no major incidents or breakdowns to deal with, although there always seemed to be a wretched Cupedo valve that needed attention, so once again I frequently found myself down in the duct keel. By the end of the trip I was heartily sick of the place.

For some reason, *Jervis Bay* was not such a happy ship as *Flinders Bay* – nobody was actually unpleasant, but the crowd just didn't seem to be so sociable. I am pretty sure, too, that the chief steward didn't approve of wives on board (there were three others as well as Astrid) and he did little to organise trips ashore or any onboard activities. There were a couple of exceptions, though, the first of which was at Christchurch, where most of the ship's company managed to go ashore to watch the British Lions play the All Blacks at rugby. As it happened, this was the only match on the Lions' 1977 tour that they actually won, although I doubt that it was anything to do with the support we gave them! It was a scrappy, low-scoring game and the Kiwis thought they had been robbed by an ill-judged penalty, so as we returned to the ship through the many disgruntled fans we kept a very low profile.

The next occasion was in Sydney, where tickets had been purchased for us to see the musical *Hair* in the famous opera house. The plot is all about the hippy culture of New York, living in the Age of Aquarius and avoiding being drafted to fight in Vietnam – it was just about bottom of the list of any productions Astrid and I might have wished to attend, but was worth it just to see the inside of the opera house itself, which was magnificent.

From Sydney we called in at Melbourne and Fremantle again before heading home. This time we went via Suez, which avoided the long drag around the Cape. The Suez Canal had not long been reopened and as we made our transit Israeli tanks and other military hardware were still much in evidence on the east bank. It was the one and only time I ever went through it. Despite the shorter distance, this trip took 79 days according to my discharge book, so there is a possibility that we visited an additional port, probably in the Mediterranean, which I have forgotten about. Either way, in just six months Astrid had now circumnavigated the world in both directions and had become quite a seasoned seafarer.

Our *Jervis Bay*, unlike her illustrious predecessor, came to a rather sad and unheroic end: she was sold to a Korean ship-breaker in 1983, but while being towed through Biscay on her final journey, the towline parted and she drifted onto the outer breakwater at Bilbao, where she broke her back, eventually being scrapped where she lay. *Flinders Bay*, on the other hand, had a long career and operated until 1996, when she was sold to an Indian ship-breaker.

14 FINAL DEEP SEA tRIP

After our two enjoyable container ship voyages, I had enough actual steam time for my chief's ticket but was still a few months short of the twenty-one-month total required, (after you have obtained a second-class certificate, you have to start the sea time again for the chief's) so it was back to Panocean for another trip. A steam ticket was always reckoned to be harder to get than a motor one, and if I had simply wanted the latter I would have had to serve only six months of the total twenty-one on motor ships, whereas for a steam ticket nine months at least had to be on steam.

Having had our leave, we were instructed to fly out in October 1977 to Antwerp to join *Anco Duchess*, which was a just bit bigger than *Post Runner* and was powered by an 8-cylinder B&W diesel of about 10,000 hp. As an aside, she had a sister ship called *Anco Duke*, which was later to suffer the most appalling tank cleaning accident in which seven men died. The tank in question had been filled with tallow, which in itself is a harmless substance but does have the property that it will absorb oxygen from the air. In this case the tank had been cleaned with boiling water, so that all the tallow residues had been washed off the sides of the tanks and were now floating about in 2 feet of hot water at the bottom of the tanks. The next part of the procedure was to allow a day or two for it all to cool down, whereupon the tallow would solidify into lumps floating on the surface, which could then be loaded into drums by crewmen working inside the tank. Apparently a man working in there had slipped and fallen (probably overcome by lack of oxygen) and the remaining casualties, which included the chief officer, were the men who went down to rescue him. What none of them had realised was that during the two-day cooling-down period the floating tallow had oxidised and the atmosphere in the tank was therefore badly deficient in oxygen. If they had used an oxygen analyser and tested it before entering, the tank could have been ventilated first and seven lives would have been saved. Whether their testing equipment was faulty or whether they simply hadn't realised the danger we shall never know, but I still find it strange that after the first few men had gone in and failed to come out again, the remainder had not put on breathing apparatus.

The earth's atmosphere normally contains about 21 per cent oxygen, and most of the rest is nitrogen; even at the top of Mount Everest, where the air is a great deal less dense than at sea level, the percentages are the same, which is why some people have managed to climb it without carrying additional oxygen. If at any altitude the oxygen percentage were to drop to about 16 per cent, however, your mental faculties would start to be impaired and you could become disorientated and confused. At 12 per cent most people will be unconscious, and at anything below 10 per cent they will die. Oxygen deficiency is insidious: the air you are breathing smells just the same and you won't realise that anything is wrong until it's too late …

Anco Duchess had a Chinese crew and was not a UMS ship, so for me it was back to watchkeeping. Once again I was the only engineer on board with a steam ticket, so all the boiler work became my responsibility, which I was actually quite pleased about as it meant I was spared some of the dirty work doing cylinder units and generator overhauls. The ship turned out to be very reliable, which is why I have very few memories of what happened during this voyage apart from the fact that we visited Shanghai in Communist China.

Initially we pretty much repeated our trip on *Post Charger* – going out via Panama and then across to Japan where we called at several ports – but then we were sent to Shanghai. When we arrived, we were not allowed to berth straight away but had to anchor in the river, where the port officials checked every last full stop on the paperwork. They were accompanied by at least ten expressionless Red Guards all carrying rifles, so it wasn't much of a welcome. While this was going on they had also demanded that everyone on board be paraded on deck and counted; it was only with some difficulty they were persuaded that at least one man had to be left in charge on the bridge and another in the engine room. It was bitterly cold, and we all stood there and shivered for at least half an hour while the guards searched the ship from top to bottom. Quite what they were looking for I have no idea, but at any rate it wasn't booze and fags as they were empty-handed when they finally departed and we were allowed to haul up the anchor and carry on to our berth.

Astrid was the only wife aboard on this trip, and we had been told beforehand by the captain that the Chinese authorities would take a dim view of anyone on a merchant ship who was just along for the ride, so for our stay in the port she became the ship's librarian. To add weight to this subterfuge, she was provided with some spare epaulettes that someone had found for her. We had no idea where these might have come from, as they were actually the sort of braid the ship's doctor might have worn on a passenger ship, with two gold bars on a red background. Anyway, these she wore with one of my navy blue uniform sweaters, and looked very smart. The ship did actually have a library, consisting of a few shelves of books in the bar, including quite a few raunchy paperbacks, which no doubt the officials would also have taken a dim view of if they had found them.

Shanghai in early 1978 wasn't a bit like the city you find there today, with its masses of new buildings and skyscrapers: what we saw would not have changed a great deal from a century before, when the first substantial western-style buildings were constructed along the part of the river's edge known as the Bund. Although Shanghai had been one of China's principal trading ports as far back as the 17th century, it was western businessmen who had wrangled the trading concessions from the local officials (where they were not subject to Chinese laws) that had developed the area around the Bund, almost as a private enclave within the city.

We stayed in Shanghai for nearly a week and everyone was able to go ashore if they wanted to; strangely perhaps, our Chinese crew were less enthusiastic about going than the officers. Every time a group of us went ashore, we were accompanied by a Communist party official who decided where we could or could not go; nobody was allowed off the ship on their own. Naturally, Astrid was very keen to see the place, and one of our first excursions was to the Chinese Friendship Shop, which was basically a huge souvenir emporium, containing all manner of goods including ceramics, carpets, wooden utensils and toys, framed pictures, and hand-beaten brass and copperware. It was all beautifully made and incredibly cheap. Periodically, English lessons were played over the loudspeakers in the store, which came as something of a surprise to us.

Unlike all the Chinese officials we had met up until then, the female shop assistants we met were friendly and smiling, and dressed in smart western-style clothes. They were desperate to sell us stuff and had probably been told that all westerners were millionaires – which alas we were not, as there were some tremendous bargains to be had. Astrid was very tempted to buy a beautiful hand-painted dinner service which was just a few pounds, but we were mindful that anything we bought would have to be carried back to the UK, quite probably by plane, so we resisted. We did buy quite a few smaller items, though, including a couple of ornate fans, boxed chopstick sets, a very nicely decorated vase, and a wooden lantern with hand-painted glass panes which came as a self-assembly kit; both the latter items miraculously survived the trip home. Astrid also bought some rolls of cloth and a couple of Chinese cheongsams. When we got home I converted the lantern to take a bayonet lamp fitting, and it currently hangs from the ceiling in one of our bedrooms.

One less desirable feature of the store was that there were metal spittoons placed all around the place, as the Chinese at the time were relentless spitters. I wondered whose unpleasant task it was to empty and clean these, and whether or not he or she received the same pay as the shop assistants. If you believe what has been written about him, Chairman Mao's first job had been to collect the night soil from around his village and spread it on the fields as fertiliser, so perhaps a similarly lowly task such as spittoon emptying was an essential requirement for anyone wishing to progress to the upper echelons of the Communist party.

Walking to and from the ship we passed through some of the less prosperous parts of the town and were able to see a little of what life was like for the ordinary Chinese citizen. For a start, this being 1978, there were no private cars on the streets, although we did see the occasional limousine on the main roads, no doubt carrying a senior party official – on the other hand, there were bicycles everywhere you looked. The buildings we passed in the back streets were mostly single-storey structures with flat roofs and front doors that opened directly into the road. Here and there we met a few small groups of people who gave us curious stares as we passed by; some were huddled around simple braziers which looked to be burning charcoal or compressed coaldust briquettes. It was midwinter in Shanghai and the weather was freezing, but we were at a loss to see what good an open-air heater such as this would do unless the people had nowhere else to go. Nearly everyone we saw, men and women alike, were wearing the ubiquitous Chairman Mao jackets and trousers in grey-blue or khaki, although a few of the women had pink jackets, which provided about the only splashes of colour. Possibly the biggest difference, however, between any western city and Shanghai was the complete absence of advertising material of any kind: if for instance you passed a grocery store, the window display might simply be piles of tins, all in white with Chinese characters in black written on them, so unless you could read them, you would be quite unable to tell whether a tin contained baked beans, bamboo shoots or bat soup.

The following day we had an organised excursion to the Shanghai carpet factory, which the officials seemed very keen to show us. Including Astrid and me, about eight of us went on this trip – but none of our Chinese crew. Before we started the proper factory tour, we were all taken to a Greeting Room, where we were sat down around a table and given a cup of hot water in which a few green tea leaves were floating – I thought that if this was what China tea is supposed to be like, then give me a cup of English builder's tea any day!

During the tea drinking one of the party officials gave us a long dissertation on the evils of the 'Gang of Four' and how they had been overthrown by the will of the people. I have to confess that we knew very little about this except that Chairman Mao's last wife had been one of the ringleaders. We were told that during the cultural revolution of 1966–76 they had slowly been taking over many aspects of state control, including most of the newspapers, as Mao's own powers declined, and had gradually been leading the party away from the 'true path'. When Mao died in 1976, the Gang of Four only lasted a few weeks longer before they were overthrown – apparently to great rejoicing on the streets of Beijing. Eventually the interminable lecture did actually come to an end, and we were allowed out into the factory – but not before we had all been ceremonially presented with a small copy of Chairman Mao's famous Little Red Book to remind us of our visit.

In the factory there wasn't a machine tool of any kind in the place. Instead, every stage of the manufacture was done by hand, with row upon row of women sitting at their looms, winding the yarns onto bobbins and trimming around the patterns with scissors to give a relief effect. The carpets themselves were wonderful and we marvelled at the skill of the workers, some of whom at least gave us a smile rather than a bleak stare. There was a shop displaying a selection of the finished products but although these were also very cheap compared to the prices we would have paid in England, they were still going to cost rather more than we could afford, and once again we had to consider how we were going to cart the things home.

Our final run ashore was a special meal to celebrate Chinese New Year, which was held in the restaurant attached to the friendship shop. The visit had been arranged by the head man of our Chinese crew, and on this occasion quite a few of the crew attended as well, so it turned out to be a very jolly occasion. The menu in the restaurant was amazing, with literally hundreds of items to choose from. Closer inspection, however, showed that it was simply the way it had been written out that was responsible for this: for example, you could order fried egg with toast, fried egg with bacon, fried egg with beans, fried egg with bacon and toast, fried egg with bacon, beans and toast, then ditto the whole lot with scrambled egg instead, and so on. Astrid had made a visit or two earlier, on her own, and had ordered a glass of rice wine. When she left, the bottle the wine had come from had been put in the fridge and had been kept for her next visit each time, and the remainder was still waiting on this last occasion too.

The following day we departed from Shanghai and were actually steaming down the river when we were once again boarded by some officials and a whole lot more stone-faced Red Guards, whose main purpose appeared to be searching the ship for stowaways. Again we all had to parade on deck to be counted, and it was even harder to convince them that even at dead slow ahead the ship still actually needed someone on the bridge to steer it and another down below to control the engine. When the guards were finally satisfied that we were all present and correct and that there were no extra people on board, they departed by launch and we could breathe a sigh of relief. Astrid and I both felt that although we were very pleased to have had the opportunity to visit Communist China, we were even happier to know that as we were despicable western imperialists we were free to go home, or indeed anywhere else we chose.

After Shanghai, the ship did the rounds of the Philippines, loading palm oil and coconut oil for our return to Europe. After leaving one port there was a problem with one of the cargo pumps, which meant me spending some time working in the pump room. I was not looking forward to this one bit, as my previous experiences in pump rooms had all been on normal tankers, where there always seemed to be a horrible stink of crude oil and everything around the pumps was covered in black filth – real treacle mines, in fact. This pump room, however, was not a bit like that – despite the age of the ship, it was still quite clean all around the pumps, while the bilges, instead of looking like the Black Sea on a dark night, contained just a few inches of yellow palm oil that had come from the leaking pump. This was much more pleasant to be working around, and the chief officer told me that it was actually very good for the skin; indeed, after I had spent a couple of days working down there, rebuilding the pump with a new mechanical seal and completing a few other outstanding jobs, my hands were as soft as the proverbial baby's bottom.

The rest of this trip was simply another long haul back across the Pacific, through Panama for my tenth (and final) time, then across the Atlantic to Liverpool, where we were paid off. I had pretty well decided by this time that I would not do any more deep sea voyages and would look for a job on the ferries instead, as we wanted to start a family. Astrid had actually been pregnant when we paid off *Jervis Bay*, but had sadly miscarried and we both agreed that as I was going to be ashore for some time while studying for my chief's ticket, this was as good a time as any to make the change from FGN (foreign going) to HT (home trade), as the type of service is described in the discharge book.

Panocean was obliged to pay me for three months' study leave and this, together with the two months' normal leave I was owed, meant that we had some time to look around and see what was going on the ferries, or elsewhere on other coastal ships. First of all, however, was the small matter of actually getting my certificate, so to start with I had to enrol at Poplar Tech once more. Dr S and Mr Burrage had both retired, so there were no familiar faces to greet me. Another difference was that the West India Docks at the back of the college were now completely deserted and there wasn't a single funnel or mast to be seen peeking over the tops of the warehouse roofs, while the railway tracks seemed to be devoid of trains, and weeds were growing up through the ballast. Poplar, too, seemed even more rundown than ever, and without the dockers to keep them open several of the High Street pubs had closed. Commuting by car up to Poplar was just as grim as ever, with long tailbacks in the mornings on the Blackwall Tunnel approach road, but this still seemed preferable to going in by train, tube and bus as I had done as a cadet.

I was going to have to pass the same four subjects, but in considerably greater detail, that I had taken for my second-class exam four years previously, namely a combined ship construction and naval architecture paper, electrotechnology, general engineering knowledge and steam engineering knowledge. Passing these, together with the oral exam, would give me a first class steam certificate. Assuming I did pass, I could then take the motor endorsement paper as soon as possible afterwards, which would give me a combined steam and motor certificate: a double-barrelled chief's ticket, as it is commonly called.

Once again I found the electrical work the hardest, nearly all of it being theoretical. In my humble opinion, very little of this would be of any use to an engineer on board ship trying to deal with something that didn't work when you pressed the button, whereas some instruction on the correct ways of testing and fault-finding definitely would have been. The guy taking us for the electrical work was quite a decent chap, though, and had a fund of funny stories which helped us through as we struggled with the boring equations, three-phase unbalanced load vector diagrams and the like. The naval architecture guy, on the other hand, was quite the reverse and his lectures were dull in the extreme; but he had some good hand-outs for us to take home and study. As for the practical papers, by now I had realised that you either knew the stuff or you didn't, and if it was the latter then it was unlikely you would be able to learn it in just three months. One thing that sustained my hopes during this period was to look back at some of the chief engineers I had sailed with and to think to myself 'If that idle bastard can get a chief's ticket, then so can I.'

After a couple of months of studying hard, I decided that I would put myself in for the next set of exams, knowing that if I failed I would have the time for at least one more shot at it before Panocean stopped paying me. As the exams drew nearer I was up until midnight most nights working through old papers and studying my lecture notes. When the actual day arrived, I left home an hour earlier than usual, and prayed there would be no stoppages at the tunnel. In the event, I arrived an hour and a half before the exam was due to start and sat in the library with my books for some last-ditch revision. The papers were three hours each; we had the two theory papers on the first day and the practical ones on the second. Once again, I found the practical papers okay, and thought I had probably passed the naval architecture, but had serious doubts about the electrical one.

After the exams were over there were a few days to wait before the oral exam, but I carried on attending lectures at the college as I didn't want to tempt fate by stopping straight away. The oral exam was not taken at Poplar but at a particularly dingy office near Aldgate, so for this I travelled up by train to London Bridge. It turned out to be a lot harder this second time around. and my three examiners seemed to have an uncanny knack of avoiding everything I was comfortable with and moving on quickly to things that had me struggling. I got a particularly tough grilling on the electrical stuff, but fortunately the questions were mostly based on practical scenarios which I was better able to answer than the theory questions.

After some 40 minutes, I was beginning to think I was doomed, but then they stopped quite abruptly and had a brief chat, before saying, 'Congratulations Mr Richardson, you have now passed your First Class Certificate for steam.' I was pretty well shaking with relief after this, and had to call in at the first pub I passed for a drink to calm myself down, before phoning Astrid to tell her the good news.

Then followed another couple of weeks at Poplar before I could take the motor endorsement paper, which I thought was pretty easy – after all, a diesel engine is still just a diesel, however big it is. Finally, it was back to the Aldgate office for another oral, which lasted barely 20 minutes. And then it really was all over – my ambition of 14 years before had been realised and now, in 1978, I had become the latest holder of a Certificate of Competency, First Class Engineer, Steamship and Motorship to give it the full title. Apart from marrying Astrid and raising three wonderful children, as I look back on my life now, I still regard it as my proudest achievement.

15 CROSS-CHANNEL FERRIES

While I had been at college I had met one of the cadets I had started with in 1964 and we were chatting about what we were going to do next. He thought he would be able to go ashore altogether and was looking for a shore-based position as a surveyor or ship superintendent. When I mentioned that I fancied going on the ferries but hadn't seen any vacancies, he asked me if I knew that P&O had started operating out of Dover and might be in need of engineers. So I phoned its main London office and made some enquiries. It turned out that although it had started with just one ship, *Lion*, on the Dover–Boulogne route in 1976, it had recently bought a second one and was in fact recruiting – it was a wonderful stroke of luck.

After the phone call, I wrote in with my CV as requested and awaited results. Just a few days later, the reply came back asking me to report to its Dover office for an interview, at which I was offered a second engineer's job on the spot, and the next stage of my career had begun.

Before I could start with P&O, I had to have a medical examination at the port health office in Dover. When I turned up for this, I was shown into a somewhat poky and dilapidated room, in which the doctor sat behind a desk, smoking a fag and reading the paper. When he eventually noticed my presence his opening question took me somewhat aback: 'Are you one of the geriatrics from Sealink or one of the skivers from Townsends?' After a moment's thought I replied, 'Actually, I'm one of the gentlemen of the P&O,' which raised just the hint of a smile. The ensuing medical was the most perfunctory I ever had – he didn't even check my blood pressure. Reading from a form he asked me whether or not I had ever suffered from any of a whole list of ailments, including the clap (his word not mine), before ticking all the 'No' boxes. Next he asked me to read a few lines from the eye test chart on the wall and finally I was shown some colour cards and asked to identify them. After a few obvious ones, he turned up a card that was a sort of bluey-green colour – or was it greeny-blue? Giving this one some extra consideration, I eventually said 'turquoise', to which he replied, 'Nobody's ever said that before,' shortly followed by 'Okay, you'll do' – and that was it!

Although P&O is now the biggest ferry company operating out of Dover, when I joined it in 1978 it was very much a minor player. *Lion* had previously been used on the Irish Sea route from Ardrossan to Belfast or Larne, while the second ship had come from Denmark and had been called *Kattegat*. This P&O renamed *Tiger* and was the vessel to which I was assigned. Of the two ships, *Tiger* was much the prettier vessel, looking very like a miniature cruise liner, while *Lion*, with her two very small funnels stuck low down on each side, was rather an ugly duckling. Both were of a similar size and could carry about 200 cars or 30 HGVs. What made the venture successful was that P&O had opted to add a little luxury to the cross-Channel experience and all the passenger decks were carpeted throughout, while the bars, restaurants and shops were very smart compared with those of Sealink and Townsend Thoresen, which up until then had had no real competition.

The Sealink ships in particular were at that time very basic, and crossing the Channel with Sealink instead of P&O was rather like choosing to fly with Ryanair instead of business class with British Airways. To add to the idea that crossing the Channel could be an enjoyable experience in its own right, P&O invented what later became known as the booze cruise, when you could go across to France on one ferry, enjoy a few hours' shopping and maybe a meal in Boulogne, before returning on a later one. This was an absolutely brilliant marketing ploy – and Boulogne was much better suited to this venture than Calais, which was firmly in the grip of Townsend and Sealink, anyway. In Boulogne the town centre shops were only a ten-minute walk from the ferry terminal, whereas at Calais you needed a bus to get into town. The crew also took advantage of this and it was common practice to send a few men ashore on shopping trips to bring back wine and cheese for their shipmates when they were about to go on leave. Another idea P&O had was the 'Dance to France', where a band or disco would be set up in the main passenger lounge and the punters could simply go to France and back for the ride while purchasing their duty-free goods in the onboard shop. These too were very popular, and increased the revenue on crossings that would otherwise have been mainly dependent on freight traffic.

All the ferry companies working out of Dover at that time operated multi-crew systems, where the crew went home in between shifts. P&O crews worked 12 hours on and 12 hours off for a week, and then had a rest week – or to put it another way, 84 hours one week and nothing the next. You would work a week of days, take a rest week, then a week of nights and another rest week and so on. To cover sickness and annual leave, this meant that P&O actually had to employ five sets of people to staff each ship. This may sound very expensive in manpower compared with a deep sea ship, but on the latter the company would still have to provide a relief crew when one lot went on leave, and would also need three sets of watchkeepers per ship, together with decent cabins and bed linen etc. Sealink, meanwhile, had a different system entirely, working its crews 12 hours on and 24 hours off for two or sometimes three weeks and then the rest week; in the end it needed about the same number of crew in total. Until I actually worked for Sealink some years later, I always thought that this system would totally screw up your body clock with alternating day and night shifts, but the Sealink employees I spoke to all thought they had the better deal and could never understand how we could manage to work a whole week of night shifts. I am not sure what Townsend did, as I never worked for them.

When I started work in Dover, Astrid and I were still living at my parents' house near Croydon, which was too far from the docks for me to be able to commute between shifts, so for my first few weeks on *Tiger*, I got a room at the East Cliff Hotel, which was pretty well the first building you would see when leaving the Eastern Docks and the bar of which was a popular venue for ferry officers at the end of their shifts. After a couple of months, the third engineer on my crew bought a house in Shepherdswell, a small village just a few miles from Dover, and to help pay the mortgage he took in the fourth and me as lodgers; this was a lot cheaper than the hotel, but also a lot less comfortable.

Joining the ship for the first time I was very impressed with the décor and fittings throughout the passenger spaces – but down below in the engine room things were not quite so smart. She had two V10 B&W four-stroke diesels driving controllable-pitch propellers, and as they had to be squeezed in below the vehicle deck there was very little headroom and because there were exhaust leaks which we never had time to fix, the top half of the engine room had a fine patina of soot everywhere. Running to a scheduled service meant that there was only enough time in port to do very basic engine maintenance, and anything that was going to take much over an hour to fix would have to wait until the next Saturday night, when one crossing was missed, giving about seven hours' working time.

I have been asked many times whether or not it was an easier life on the ferries compared to deep sea and the answer is: it was different. If you were on a ship in the middle of the ocean and you broke down, you simply drifted around until you fixed it. On the ferries, there was always the relentless pressure of the schedule to deal with: every six hours, day or night, your ship was expected to sail from Dover and if you had leaking exhausts or one of the engines was only running on nine cylinders or had any number of other minor defects, then you just carried on as best you could; only on the Saturday night layovers was there a chance to do any of the bigger jobs. *Tiger* was not as fast as *Lion*, so we had less time in port anyway, and all you could really expect to do in a normal turn round was to change a main engine injector – or maybe two if the chief was also prepared to get his hands dirty.

Because we were slower than *Lion*, it meant we had to work our engines harder to maintain the schedule, and for quite a lot of the time they were going absolutely flat out. The limiting factor was the temperature of the exhaust gas leaving each cylinder: if you ran continually with temperatures much above 500°C, then before long the exhaust valves would burn out. Each cylinder had its own temperature probe, which led to gauges in the engine control room; these gauges probably got more scrutiny on a crossing than all the rest put together!

There are several reasons why one exhaust might be showing a higher temperature than its neighbour, and it might just be that the thermometer itself was faulty, so if we did have a hot cylinder the first ploy would be to change the temperature probe or swap it with the one beside it and see what happened. If it still showed hot, we would change the injector for an overhauled spare at the first opportunity – and if that didn't work either, then it would mean that one of the exhaust valves was on its way out and would also have to be changed. Unlike a car, which would require the cylinder head to be removed to do this job, on *Lion* and *Tiger* the exhaust valves came as assemblies which included the valve itself and a spring built into a housing as one unit; this was bolted down into a pocket in the top of the cylinder head and could be removed the same way. The rocker gear had to come off first, of course, and the cooling water drained down from that cylinder before you started (each one had individual shut-off valves) – but if the gods were with us we would just have enough time to change one in a turn round at Dover. Occasionally one of them would absolutely refuse to come out, due to carbon from the exhaust gas having leaked past the seals and jamming it in the pocket. We had no option then but to bolt it back down again and shut off the fuel to that cylinder, so from then on the engine would be running on nine out of ten cylinders, slowing us down even more. We would have to continue like that until the next Saturday night layover, when we could change the whole cylinder head and bash the reluctant valve out from the inside.

All the ferries berthed stern first in Dover and would not therefore have to be turned around when departing, which meant that if we hadn't finished the job and the weather was okay, the captain would be able to sail on one engine. When I became chief engineer and had any engine job that was likely to overrun the turn round time, I would always ask the captain if he was happy to do this, on my assurance that we would give him the second engine back before reaching Boulogne – and we always did, although it was a close-run thing on more than one occasion.

The deck machinery could also cause us a lot of stress at times. Limping along with engine problems at reduced speed was one thing, but if you ever arrived in port and were unable to open the bow or stern door to let the vehicles off there was a real panic, as it didn't take very long for the drivers to get impatient. Despite being told not to start their engines until they could see their lane moving off, many of them did just that, and this rapidly filled the car deck with noxious fumes. At the front end of the car deck there was a visor which hinged upwards, very much like the one on the helmet on a suit of armour, and behind this was the actual bow door, which was then lowered like a drawbridge. At the other end, the stern door consisted simply of a ramp that opened in the same way as the bow door. Neither the bow nor stern ramps were lowered until the vessel was actually tied up on the berth, but the visor had to be lifted beforehand as it would otherwise have fouled the shoreside loading ramp, or link-span, as it was known. The visor was held down and locked in place by hydraulically operated pins, while the two doors had a series of clamps (or dogs, as we called them) around the edges. To avoid damaging any of the hydraulics by trying to move anything before the dogs were released, each one was fitted with a microswitch which would be closed when the dog was in the off position, and all of them had to be closed before anything would work. So upon receiving the phone call from the bridge to say they couldn't open one of doors, the electrician and an engineer would be sent forth to clamber all around it to find the faulty switch. Once identified, it might only need a wiggle and some WD40 sprayed over it to get it to work – but if it was going to take any length of time to fix the electrician might simply short it out until it could be swapped for a spare.

If the defect wasn't a faulty switch, the next most common problem was a burst hydraulic hose, and although we carried a good stock of spares it was not a five-minute job to change one, in which case tempers on the vehicle deck could get very frayed indeed. I don't know whether these hoses had a built-in life (or death!) expectancy, but Ardee Hoses, the firm that supplied us, must have been pretty well kept in business by P&O at that time.

The vehicle deck on both ships was high enough for an HGV, but if we were only carrying cars the available space for them could be increased by lowering mezzanine decks with loading ramps, which would enable us to transport the cars on two levels. When the mezz decks, as we called them, were filled, the loading ramps would be raised and the second lot of cars driven in under them. When unloading, all the drivers sitting in their cars on the mezz decks would have to wait until the bottom layer had been discharged before getting their turn to drive off. The booking office tried to arrange that as many as possible of the HGVs travelled on the night crossings, to free up more space for the car traffic during the day, so the mezz decks were seldom used at night. None of this concerned the engineers at all, except when a hydraulic hose failed and left a mezz deck stuck halfway, in which case it suddenly became the no.1 priority.

One other all-important factor in ferry operation is the weather, and for my first winter in Dover it was vile, with winds of gale force 8 or more for what seemed like months. Apart from being uncomfortable for the passengers, it played havoc with our shift system. In high winds ships would quite often struggle to berth, possibly having to take a tug to assist them, while the others queued up outside the harbour waiting their turn, which of course meant that your nominal 12-hour shift quite often finished up being 13 or 14 hours. I can remember one particularly dire occasion when the port had to be closed and we sailed around the corner past Deal to ride out the weather before finally getting off the ship some 19 hours after joining. On that occasion, the company cancelled a crossing to get us somewhere near back on schedule, which meant our nominal 12-hour rest period was reduced to around 6. As there were no mobile phones you could not contact anyone on board to see how late the ship was likely to be, and trying to get information from the company office was pointless as they would always tell you that as far as they knew the ship would be running on time. This meant that you just had to turn up at the appointed hour and then sit around in the terminal building waiting for it.

The engineer officers in particular never got off the ship on time anyway, as it was usual to chat with your relief for a while before handing over, to discuss any ongoing engine problems. And in any case when the ship actually arrived you might still be in your boiler suit and need to get cleaned up and changed. If the ship was only an hour or so late the oncoming shift might then be able to catch up the lost time by making extra quick turn rounds, so any extra hours you had worked would mean that many hours' less rest period before you had to turn up again for the next shift. I suppose that in the end it was a case of swings and roundabouts, and that for every extended shift you worked, then sooner or later you would get a shorter one to make up for it; but the overall effect was very tiring, and after a week of nights in particular you could be pretty well exhausted.

The ships themselves also suffered from the weather, getting dents and damage as they struggled to enter the harbour and get tied up in the gale force winds. P&O

operated from nos 1 and 2 berths on the far side of Dover Eastern Docks, and to approach them the ship had to come in stern first through the gap between the jetty that ran out from the berths and the old submarine pens. These were set at right angles to the eastern arm of the harbour and extended out toward it. During her first year of service *Lion* clouted the pens on several occasions, prompting the sentiment that she had done them more damage than the Luftwaffe had achieved during the war.

Some captains would steam in through the eastern entrance of the harbour, turn to port and carry on across until they judged they were in the right position to reverse all the way back into the berth. *Lion* quite often used this method of approach and employed her bow rudder to assist her when steering backwards. On *Tiger*, however, most captains preferred to enter the harbour forwards, then head straight for the berth, before bringing the ship to a stop and swinging round on the spot by using the engines and bow thruster, until the stern was pointing the right way. This 'swinging off the berth' method saved a little time but had the disadvantage that if the wind caught the ship at the wrong moment, it could cause it to drift down sideways onto the end of the aforementioned jetty.

Captain A, who was senior captain on *Tiger*, did just that one day: I was chief engineer by then and was sitting in the engine control room on the port side of the ship with the lecky and the third engineer. The engines seemed to be getting an even bigger thrashing than usual, then suddenly there was an almighty thump which almost tipped us out of our chairs: we had struck the pier at a point on the ship's side right next to the main switchboard. In the control room, all the instruments and controls were on the inboard side, with windows looking out over the engines, while behind us was the switchboard with a narrow walkway behind it. When we went back there and had a look we could see a big dent in the ship's side, and a couple of the frames were buckled too. Nothing was leaking, though, so when the ship finally got tied up and the captain phoned down for a damage report I was able to tell him that we were still good to go. On the very next trip exactly same thing happened, and the captain actually managed to put another dent right on top of the first one! This time we were not so lucky: sea water was spurting in from several cracks and spraying the back of the switchboard, which was somewhat alarming. We swathed everything in rubber sheet to keep the water away from the electrics and telephoned the bridge to say that we would need to go to dry dock for repairs, so after discharging our passengers and all the spare crew, we sailed off to Le Havre to get it fixed.

I continued living in my digs for this first winter, but as the service seemed to be doing well and the job secure, Astrid and I decided we would finally sell up my parents' house and move somewhere closer to Dover. Astrid did all the house-hunting and eventually found a three-bedroom bungalow in a village called Stelling Minnis, about 8 miles due south of Canterbury. My first look at the property was in the pitch dark, just having finished a day shift, but it was obvious that not only was it a cosy dwelling but also it was in a very quiet rural area, which we had both wanted. Stelling Minnis at that time had no street lights, no mains gas and no mains drainage, and cows were frequently to be seen strolling down the high street from one bit of grazing to another. Nevertheless, it did have a village shop, a pub and a post office, so we agreed a price and moved down there in March 1979. Getting to Dover from Stelling Minnis was actually not that easy and meant a trip of about 18 miles through some quite narrow but also very pretty country lanes, which became my preferred route, but if the weather was really bad there were a few alternatives on better roads at the expense of two or three extra miles each way. Most of the crew lived in and around Dover or the nearby villages, with Whitfield being quite popular; despite the fact that this gave them an extra hour at home between shifts, we were very happy to stay out in the sticks at Stelling Minnis, and we remained there for the next 14 years.

In February 1980, P&O started a third ship on the service, this being *Tiger*'s sister ship, which it had also purchased from Denmark and renamed *Panther*. Several senior officers from *Lion* and *Tiger* were transferred to her, which meant promotion opportunities for those remaining, and I was immediately made chief engineer – a position I held until 1985 when P&O decided to sell us out to European Ferries, which owned Townsend Thoresen.

I had been in the job barely a few weeks when I joined the ship after my rest week to find the off-going chief in a state of doom and gloom and still in his boiler suit. It turned out that one cylinder had developed a low exhaust temperature for a change, and over the previous two days they had tried various things to fix it without success. After proving the temperature probe was okay, they had changed the injector, which also made no difference, so then they thought the fuel pump must have somehow packed up and had changed that as well – again, it made no difference. Next up, someone had the idea to check the compression on that cylinder and they found there was virtually none at all. This then led them to believe that perhaps a lump of one of the exhaust valve heads must have broken off, so they changed both of them – which once again had no effect. This was the situation when I joined, and the thinking on board was that there must be a hole in the piston itself, which would mean a big work-up at the next Saturday night layover to remove the cylinder head in order to change the piston.

When we started the engines for departure, I got one of my engineers to take off the rocker cover on the offending unit so I could have a look at the valve gear, and it only took a few seconds to see that the timing was obviously way out. In a four-stroke engine, the exhaust valve opens first, then just before it closes the inlet opens; then there is a pause while the compression and power strokes take place. So the rhythm goes: exhaust, inlet, pause, exhaust, inlet, pause etc. This set, however, was going exhaust–inlet, exhaust–inlet with no pause in between, which meant there was always one valve at least partly open, letting the compression escape. The next step was to find out how this could have happened, so we removed the camshaft cover to have a look in there too. On these engines the cams were made individually and then secured in place on the shaft hydraulically, using finely tapered sleeves in the same way as tail shaft couplings, and we could see immediately that one of the cams had shifted well away from its proper position. We still could not fix it, however, as it required some special tools and a hydraulic pump which we didn't have, so I contacted our office to arrange for a maker's man to visit the ship and do the job for us, while in the meantime we carried on running on nine cylinders. You had to feel sorry for the previous crews who had done all that extra work for nothing, while we had found the problem without even getting our hands dirty!

The B&W engines on *Tiger* and *Panther* were not exactly a popular design; indeed, I never knew of any other ships that used them. Any marine diesel will knock a little when starting up, but the racket these two V10s made would have woken the dead. Imagine the sound you might get if you were hitting a hardwood railway sleeper end on with a 14-pound sledgehammer, and that will give you some idea of the noise they made. Upon putting the lever over to the start position there would be the usual hissing and shoofing noise as the air went on, but as soon as you went to the run position it would go something like this – donk, donk, kadonk, kadonka donk, kadonk, donk, donk, donk, donk – before finally settling into the usual grumbling tickover noise. Someone starting the engines up for the first time would usually take a step back in shock and ask the nearest person if they always sounded like that.

While P&O would happily pay almost any amount to keep the passenger spaces looking perfect, when it came to paying for engine spares the company was as tight as the proverbial duck's rear end. The nice carpets it had fitted to the ships originally had proved to be difficult to keep clean when up against seasick passengers and cigarette burns, so they had been replaced with Flotex, which even back then was one of the most expensive industrial-use carpets you could buy, so the bill for the three ships must have been enormous. Down in the treacle mine, what we wanted more than anything was a complete set of new injectors, which would only have cost a fraction of what P&O had spent on the carpet, and yet all we ever got were overhauled spares. There is a limit to how many times the valve and seat of an injector can be reground before the performance starts to decay and we had passed this point, so that every time we started the engines a black pall of smoke issued from the funnel, and if the wind was blowing the wrong way this would drift across the passenger terminal. One day I was in my office prior to departure when the superintendent came storming down the stairs and started berating me for making smoke, even asking if we had changed over to burning coal! I simply showed him all the previous spares requests, with the note 'not received' written alongside them. The new carpet, incidentally, was light brown with dark brown spots, and was immediately dubbed 'fag burn pattern'.

When I had been promoted to chief engineer, I had determined that I would not be like so many of the others I had sailed with, who had been either bone idle, alcoholic or both, so whenever we had a work-up in the engine room, I would always go down to lend a hand, or simply mind the shop in the control room if my crew were managing okay. As there were only three other engineers and a lecky on a crew anyway, an extra pair of hands was nearly always appreciated, and I don't think any of the seconds I sailed with ever resented me being down there. My quarters were situated below the vehicle deck and consisted of a small office which led into a substantial dayroom with a three-piece suite, and a bedroom with toilet and shower. I can honestly say that during the three years I spent, night or day, as chief on that ship, I never once slept in that bed.

On one occasion we had a piston to change, and after spending an hour or so to help get the job started, I went back up to my office to do some paperwork. Some time, later John Pett, my very competent second at the time, stuck his head around the door and said he had been struggling for half an hour and had been unable to get the piston back down into the cylinder, and would I mind having a look? With any Vee engine, whenever you removed a cylinder head or piston, it was never a straight lift, so two chain blocks would be required to hold things at the correct angle. When I went down a few minutes later and walked across in front of the engines, I could see straight away that the piston was not quite lined up with the bore, while John, who had been working at the side of the engine, would not have been in such a good position to judge the angle. It took me just a minute or two to adjust the chain blocks, after which I gave the piston a good shove with my boot and it slipped straight in. I don't think he ever forgave me.

My regular fellow officers for my first two years as chief consisted of Captain Frank Quick, who was a very good ship handler and had earned the soubriquet 'Quick by name and quick by nature'. We sometimes shared a car into work, as he lived just a few miles away from Stelling Minnis. I always thought he had the better half of this arrangement, as he would be whisked down to the docks in my E-type, while I had to put up with his slow and noisy Land Rover. Another captain of note was Mike Forwood, an imposing figure with a thick black beard who could be quite fierce at times – behind his back he was always known as Ahab. He was a proper sailor, and after leaving P&O went on to become master of the well-known sail training ships *Winston Churchill* and *Malcolm Miller* amongst others. I would have gone anywhere with him.

I had several different seconds, starting with Dave Clark, who had a fund of interesting stories about the time he had spent in East Africa on the Lake Victoria steamers. He was always known as 'the real Dave Clark' as opposed to his namesake in the company, who was simply 'the other Dave Clark'. After Dave, I had the above-mentioned John Pett, who sadly died in 2019, and then Tony Warne, a quietly spoken and intelligent officer who I thought would go far – and did, later becoming a senior chief engineer with P&O.

My regular third was a guy called Norman Oakley, who, even though he was a conscientious worker, somehow always managed to look dirty and scruffy. He loved to find something to take to bits – and even if it was just a ten-minute job he would usually manage to get himself filthy in the process; it was as if he had been fitted with a personal mechanical lubricator to anoint himself with oil. After I had been chief for a few months, I thought I would try and get the engineers' changing room tidied up, as it had piles of discarded boiler suits and mouldy old boots left there by previous and long-departed crew members. I duly put up a notice to say that anything still lying around in a month's time would be binned – this gave all the crews a chance to see it. When the appointed day came, I gathered everything up and dumped the lot in the waste bin on the car deck. At the end of the shift Norman came to see me and asked if I had seen his boots. When I reminded him about the notice he replied 'Yes – but those were my going-home boots, not my engine room ones!'

Steve Humphries was the fourth engineer with whom I had shared digs in Shep-herdswell. He had the gift of the gab and was very much the ladies' man – whenever

I met him off the ship, he was usually accompanied by some very attractive female. Steve was never going to progress very far up the engineering ladder, not because he was no good but simply that his heart wasn't really in it. The last I heard of him he had become the manager of a very successful paint supply company.

Another fourth engineer I sailed with quite regularly was Rob Caldwell, who was also one of the refit crew when *Tiger* went to Palmer's shipyard on the Tyne for her annual dry dock, circa 1981. This was the first time I had visited Geordieland, where so many of Esso's engineers had come from, and I developed an affection for both the countryside and the people of the north-east that endures to this day. We were billeted in a hotel on Ocean Road in South Shields, famous for the large number of Indian restaurants spread along it; by the time the refit was over we had visited at least ten of them as well as a good number of very friendly Tyneside pubs.

Last but not least were the leckies: my first regular was an ex-RN guy called Pat Lockhart, an enormous, friendly and quietly spoken man who would come into work, whatever the weather, wearing an old navy blue raincoat and riding a little Vespa scooter – how that poor machine managed to carry his great weight I will never know. On one occasion there had been a very heavy overnight snowfall and Pat, seeing a snowdrift in front of him, had tried to charge it at full speed, using his momentum to carry him through. Unfortunately, buried under the snow was a car, which brought him to a sudden and unplanned halt, so he finished up in hospital. When P&O sold its Dover ferry business, Pat took the redundancy money and retired to the Norfolk Broads, but sadly died a few years later. Last but not least was Neil Farquhar, who apart from being a very good lecky, was a really decent man to know; he eventually became a shore-based superintendent and at the time of writing was still with P&O.

As for engine room ratings, we only had one per crew, and mine was a real old salt who had survived being torpedoed in World War II. He made us tea at regular intervals and did all the cleaning and painting throughout the machinery spaces – and apart from the soot-blackened area above the cylinder heads he made a pretty good job of it, too. For three out of the four crossings, he would work away quietly and steadily, but on the last trip he would go up to get changed when halfway across and would somehow manage to get drunk as a skunk by the time he came down for stand-by in Dover half an hour later. I could have taken disciplinary action over this, but in view of his wartime service and the fact that he was otherwise a good hand, I would always turn a blind eye, and once or twice even gave him a lift home when he was having trouble walking.

In 1982 I was transferred to *Lion*, which also had her share of problems, although the engines themselves, which were V12 versions of the popular Pielstick PC2, were much more reliable, and made a lot less noise and smoke when starting up. On *Lion*, it was the generators that had been one main cause of problems, as they occasionally self-destructed in quite spectacular fashion, with connecting rods coming through the crankcase and other such excitements. The original generator engines had been replaced at some time by Rolls Royce V8 diesels which were more powerful, and the main cause of the accidents was that these engines had originally been designed to run at 1,500 rpm to drive a 50 Hz alternator. *Lion*, however, was a 60 Hz ship, which meant that the engines now had to run at 1,800 rpm, and despite assurances from Rolls Royce that this would be okay it obviously wasn't. After about three emergency engine changes, Rolls Royce made some modifications, including upgraded exhaust valves, which were the main cause of the failures, and after that things settled down.

The boilers, too, were a persistent cause of trouble, demanding a great many hours to keep them going. *Tiger* had simple hot water boilers which were so reliable that after all this time I can't even remember where they actually were in the engine room – but *Lion* had two Clayton steam generators, one big and one small, both equally unreliable. A steam generator, as opposed to a boiler, consists of a vertical coil of steel tube surrounding the burner, through which water is continuously pumped in at the bottom end. Some of the water will be turned into steam on the way through the coil, and the mixture of steam and water emanating from the top end is taken to a separator vessel, where the water gravitates to the bottom and the steam is taken off the top. In the separator, a level had to be maintained at all times over a quite a narrow range, and the automatic controls for this were always playing up. If the level did drop too far the boiler would cut out, and there were also cut-outs that operated for flame failure and excess coil temperature (and probably one or two more reasons that I have forgotten). On some shifts we had to keep someone almost permanently in attendance to keep the wretched things going. Eventually, even our tight-fisted managers agreed that we had to get something better, and I was given the task of finding a suitable package boiler and writing a specification for its installation. The boiler finally arrived when *Lion* was sent off to refit at Immingham in 1984/5 – but by then P&O had sold the ship out from under us, so we never did get to see how well it performed in service.

Apart from the Dover to Boulogne route, P&O also operated out of Southampton, across the Irish Sea and up in Scotland, and it seemed that some of these other routes were not doing terribly well. By the middle of 1984 there were rumours that some services would be withdrawn at the end of the summer season, which would obviously mean job losses. At this point *Lion* was visited by Jeffrey Stirling, the boss of the entire P&O group. Having got all the officers and crew assembled in one of the passenger lounges, he went on to give us a pep talk, saying how well we had been doing and that the Dover–Boulogne operation was the one part of P&O ferries that didn't have a question mark hanging over it. After he had finished he did a tour of the ship including the engine room, where he cracked his bald head on a steel door frame – if we had known what was in store for us, I am sure that most of us wouldn't have minded if he'd hit it a bit harder.

After this we all carried on as usual for the next few months, feeling that our long-term job prospects were assured. We hadn't really needed any encouragement, anyway, because in the main we were proud of our little navy in Dover which, with its three second-hand ships, had wiped the smiles off the faces of Sealink and Townsend and single-handedly created the idea that cross-Channel travel could be more than just a way of getting to the other side. Imagine, then, the complete shock that came just a few months later, when P&O announced that it had sold its ferry operations in Dover and Southampton to European Ferries, which operated the Townsend Thoresen ships. Over the next few months this led to hundreds of job losses.

We did not know at the time that this was simply a means by which P&O could divest itself from Normandy Ferries, which had been its French associate when it had started up in Dover. Indeed, when I had joined *Tiger* she had had 'Normandy Ferries' painted on her sides, not P&O. The Southampton–Le Havre service had two ships, *Dragon* and *Leopard*, the latter being French-crewed. This service, in particular *Leopard*, whose crews enjoyed much better conditions and wages than those on *Dragon*, had been losing money, so P&O wanted to be rid of it. We little knew that P&O already had its eye on the bigger picture and that in two years' time it would be back in Dover as the biggest operator. Whoever was in charge of the honours list at the time must have thought it was a brilliant plan, because in the following year Jeffrey Sterling was knighted for his services to the shipping industry.

Initially, nothing much changed and *Lion* was sent off to refit in Immingham as planned. When we arrived and were settled in the dock, nearly all the crew including the captain went home, leaving just me, Brian Smith the second engineer, Pat Lockhart the lecky, and a second mate to run the show. We were on shore power, and as nobody was working a night shift we didn't need anybody to stay on board at night, so every evening around 5 pm we all piled into my car (not the E-type, which had been sold by this time), which I had brought up on the ship with me, and went back to the hotel where we had been billeted. During our stay we learnt that *Tiger* and *Panther* were to be retained and repainted with Townsend's horrible orange colours, while *Lion* was to be sold to a Greek operator following the refit. Our ships' hulls had all been painted in a very pleasant light blue and I thought the new livery and logos completely spoiled them.

It seemed that *Lion*'s refit had been well and truly put on the back burner, and work progressed very slowly, with none of the usual haste to get the job finished and the ship back in service. We did have a visit from some of the office top brass and the representatives of the new owners, but other than that we were left very much alone. To help pass the time I decided to organise a scrap drive, and my second and I scoured the engine room and stores for any scrap bronze valve bodies, pump impellers and the like. The lecky found some scrap copper cable to add to the heap, and even the second mate managed to find a few bits and pieces, so in the end we had what we thought would be enough to keep us in beer money during our stay. At this point I also had to involve Dave X, the Immingham ship manager, as he would have the contacts and the means to dispose of the stuff ashore. While we were on the car deck looking at the scrap pile, his eye happened to notice the eight spare propeller blades that were bolted up along the ship's side. A controllable-pitch propeller is not cast in one piece like a conventional one but has a hub which houses the blade operating gear, while the blades themselves are all bolted on separately. To stop the operating oil from escaping, each blade has a circular foot that bolts into a recess on the hub and is sealed with O rings. Over time, these circular feet would wear around the O-ring grooves, and oil would start leaking out, hence the need for the spare blades. None of our spares were actually new and they would all have had to be rebuilt before they were fit for use, so at the time they were in fact just so much scrap metal. Seeing the way Dave was looking, I indicated the four most worn blades but otherwise said nothing. The following morning, the scrap pile, including those four blades, had gone and we all got handed rather more cash than we had been expecting.

Lion eventually went out to the Mediterranean and operated under three different names until 2004, when she was finally sent to Bangladesh for scrapping – a very respectable 38-year working life. Meanwhile, Townsend had finally decided what to do with us, and the senior officers were summoned to a meeting in Dover, where our futures were outlined: the senior captains and chief engineers would keep their jobs and their rank if they didn't want to take redundancy, but those who. like me, had joined the company a bit later had the choice of either staying on in a lower rank or taking the redundancy money and leaving.

This was one of those times in your life when you come to a fork in the road and there really is no signpost to say where each path will lead you. On the one hand, staying on with Townsend looked as if it might have the best long-term future – but after what we had been told just a few months before by someone who was about to become a knight of the realm I was now taking any management statements with a very large pinch of salt. Additionally, I didn't really fancy going back down to second engineer again on my own ship. On the other hand, the redundancy money was substantial and I thought that there was a good chance that I could get a second's job in Sealink anyway. By now Astrid and I had two children with a third on the way, so we were thinking of adding an extension to our house in Stelling Minnis; the redundancy money would pretty well pay for it, so that was what I eventually decided to do.

Sealink of course, was the oldest ferry operator in the port and its history went back to the steam packet boats of the mid-19th century, when the South Eastern Railway Company had operated services first from Folkestone to Boulogne and a few years later from Dover to Calais – the latter route, being some 4 miles shorter, then became the most popular. The term 'packet boat' refers to the fact that one of the main functions of these ships was to carry packages of mail from England to France and back. The ships had always operated from the Western Docks in Dover under an agreement with the Admiralty, which had developed the port and built a repair yard there; this facility was forever afterwards known as the 'Packet Yard'. The headquarters of the company was in Southern House, an imposing building at the entrance to the Western Docks, which had formerly been the prestigious Lord Warden Hotel.

After making the decision to take the money and run, I lost no time in applying to Sealink for a job, and before long was summoned to Southern House for an interview. This was quite a short affair, at which I was told that Sealink would be very happy to have me but that I could only be taken on as a third engineer, which apparently was company policy. I think they must have seen the disappointment on my face for they were quick to point out that provided I got a favourable report from the chief engineer after my first few weeks' service, I could expect to be sailing as second again within months. This sounded much better, so I accepted the offer and signed on as third engineer on MV *Hengist*, working Folkestone–Boulogne.

I can't say I particularly enjoyed my first two weeks of 12 hours on and 24 hours off as it did indeed play hell with my body clock. It did at least, however, mean that in the event of delays caused by bad weather I would have enough time at home for a proper rest. The human body is very good at adapting to changing circumstances, however, and before long I settled down into the routine.

I don't know what the chief engineer had said about me after this first spell of duty, but very much to my surprise I found that when I returned after my first rest week I had already been made up to second engineer, and I never went back down to third again. This was exactly what I had been hoping for, so I was well pleased with the way things had turned out. Even better was to follow: after just a few months on *Hengist*, one of the chief engineers went off sick for several weeks and as I was the only engineer available with a chief's ticket I found myself promoted once again. I was delighted, and was also convinced that I had made the right decision over the redundancy. As things turned out, however, it also led to a very unsettled career in the following years – but more of that later.

This rapid rise through the ranks caused a lot of raised eyebrows amongst some of the long-standing second engineers, who probably thought they should have got the job instead. As I was quick to point out to anyone who questioned it, the ship's passenger certificate required that at least one engineer officer should have a chief's ticket and if they hadn't got one then they shouldn't complain. This pattern of promotion on an 'as required' basis continued all through my time with Sealink, and I probably spent almost as much time in the chief's chair as I did as second. My discharge book, however, had me resolutely down as third, and it was only when I was about to leave the company and complained to one of the captains that the entries were amended to show I had been actually been serving as second engineer – there was never any mention at all, however, that I had regularly sailed as chief engineer.

Hengist and her sister ship *Horsa* were excellent ships, and I very much enjoyed the time I served on them. They were a bit bigger than *Tiger* and *Panther,* and could each carry about 250 cars or 38 HGVs. Each powered by a pair of reliable V16 Pielstick engines, they were also quite a bit faster and were ideally suited to operations out of Dover and Folkestone. Compared to the sort of life we had had in P&O, where there always seemed to be something that badly needed fixing, they were almost boringly reliable and just seemed to keep going backwards and forwards without any fuss. In January 1986 I went with *Hengist* to refit at Swan Hunter's yard on the Tyne, and was able to enjoy another spell in Geordieland and the Ocean Road curry houses.

Another thing I liked about working on the H boats, as they were called, was running out of Folkestone instead of Dover. It was nearer to Stelling Minnis by at least 6 miles, and on a few occasions I actually cycled to and from the ship in an attempt to get my weight down. It was a very pleasant ride in the early mornings when I worked a day shift, coming down through the villages of Lyminge and Etchinghill – but returning in the evening up the gradients was not nearly so much fun. Folkestone Harbour was tiny compared to Dover, and the ships tied up directly to the inside of the pier, where a link-span had been installed to get the vehicles up onto the quay. Trains still occasionally worked down the branch line to the harbour station, which was right alongside us. This station, which had opened in 1856, crosses the harbour on an impressive viaduct before heading up to Folkestone East junction on a fearsome gradient of 1 in 30. In the final years of steam, ex-GWR 0-6-0 pannier tanks were used for banking, and it was not uncommon to have three of them pushing a train at the rear while the big Pacific at the head slipped and slithered its way up the hill. Unfortunately, this had been one of the many places I never reached in my trainspotting days and the only reminders of the spectacle are a few clips on YouTube. This line was formally closed in 2014, but the station has been restored and you can walk across the viaduct.

One of the chief engineers I sailed with on *Hengist* was a chap called Dave Giddy who, like me, was a lifelong steam fanatic. He had a boat of about 18 feet in length that he had equipped with a home-made steam engine and boiler. Periodically he would take his crew out for a trip on the Kentish Stour. Starting from Fordwich near Canterbury, we would head downstream to the Dog and Duck Inn at Pluck's Gutter for lunch before returning, which made a very enjoyable day out. I must have impressed him with my enthusiasm for steam, because toward the end of the year he asked me if I fancied a few days as chief engineer on the paddle steamer *Kingswear Castle*, whose restoration he had been involved in – an offer which I was of course delighted to accept. The vessel had only just returned to service and was running passenger trips up and down the Medway. It transpired that someone was needed on board with a chief's steam ticket to satisfy the requirements of the insurance, which meant I was really only going to be the figurehead chief while her regular volunteers actually ran the show. This didn't bother me in the least, as it was only fair that the chaps who had done all the hard work should enjoy the fun now that she was actually running again. Nevertheless, I still got to have a go at manoeuvring the compound diagonal paddle engine and stoking the boiler from time to time, so I thoroughly enjoyed myself.

6 March 1987 is a date that anyone working on the ferries at that time will remember, it being the day that *Herald of Free Enterprise* capsized when leaving Zeebrugge, killing 193 people. The disaster occurred because the ship had sailed without closing the bow doors, and upon her reaching a certain speed the bow wave had become high enough to flood her vehicle deck. At the official inquiry, three people got the blame: the assistant bosun, whose job it was to close the doors, the chief officer, who should have reported them closed to the bridge, and the captain, for not checking they had in fact been closed.

The directors of P&O by this time had a controlling interest in European Ferries, which owned Townsend Thoresen, and they were also heavily censured for a whole range of managerial deficiencies. Not least of these was that their attention had been drawn to the problem of sailing with the doors open by one of the other captains as far back as 1985 – and again in 1986 – with the suggestion that warning lights be fitted on the bridge. The company replied that it did not think it necessary for warning lights to be fitted when the man at the controls could see for himself – this despite the fact there had been two previous occasions when a sister ship of *Herald* had sailed with them open. On any ferry with a bow visor (like *Lion*, *Tiger* and *Panther*) the bridge would not need to be informed whether or not the door was shut because if it wasn't they would still have the visor sticking up in front of them as a huge visual reminder. Even so, it was the policy on those ships that whoever had shut the doors (usually the second mate) would telephone the bridge and report that the ship was ready for sea when he had closed them. Unfortunately, on *Herald* the bow doors were of the clamshell type that opened sideways, closely following the hull, and unless you leaned right out from the bridge wings, you could not see whether they were open or shut, especially at night.

There is one other point concerning this disaster which did not appear in the official inquiry and which I think should have been mentioned: this was that the ship had left the berth at 1805 and it was not until 1825 that water was first noticed entering the ship. The ship was fully capsized just a few minutes later: the bridge clock had stopped at 1828. This means that for 20 minutes there was no crew member on the vehicle deck; if there had been he would surely have reported the open doors. On all my previous ferries, there was always a seaman on the vehicle deck during a crossing whose job it was to put chocks under the wheels of any car whose owner had forgotten to apply the handbrake properly (cars were not chained down), check the lashings on the HGVs and carry out similar duties. For the most part he would probably be sitting down somewhere reading the paper, but at least he would have been there and would have made the vital call to the bridge about the open doors. On the *Herald* on that fatal day, it was plain that there had been no crew member on the vehicle deck at all.

This was the end of Townsend Thoresen as a brand, because it would forever be tainted by the disaster – and by then the P&O group, as explained previously, was the actual owner anyway. P&O lost no time in distancing itself from the Townsend name, and all the remaining ships were quickly repainted in P&O colours with dark blue hulls and the P&O flag on the funnel. P&O behaved disgracefully in the aftermath of the disaster, doing everything in its power to deny responsibility. Most of the crew of the *Herald* eventually lost their jobs. I was disgusted by P&O's attitude and I was glad I had moved to Sealink in 1985 – not least because at the time of the disaster, it had already sold *Tiger* and *Panther* and I might actually have been on *Herald*, whose chief engineer, second engineer and lecky were among those who lost their lives.

Around May 1987, I was transferred to a ship called *Cambridge Ferry*, which I think may well have been the last train ferry to work regularly out of the original ferry dock in Dover, and was running to Dunkirk. The romantic days of the Night Ferry – when well-heeled passengers could board their sleeping car at London's Victoria Station in the evening and disembark at the Gare du Nord in Paris in time for breakfast – had long gone, and all we ever carried were humble freight trains. We did have the facility to carry a few cars on the poop deck, but during my time with her this never happened.

The trains were loaded at the stern, where two tracks led onto the train deck, each train would then meet a set of points where it split; in total we could carry four trains. Despite all the fuss about fitting doors-open warning lights on ferries, *Cambridge Ferry* had a hole at the back end big enough to drive two trains through with no door at all! What she did have, though, was an island that went down the enclosed length of the train deck – thereby giving a longitudinal division to reduce the free surface effect – and very large freeing ports to allow any inrush of sea water to rapidly drain away. The central island was the only way up into the accommodation or down to the engine room, and the doors into it were watertight and could be dogged shut. There was no bow door. This arrangement worked fine, even when she was running the often stormy Irish Sea route.

Cambridge Ferry, built in 1963, had served on various routes including Harwich–Zeebrugge and Holyhead–Dun Laoghaire before coming to Dover. She was powered by a pair of 7-cylinder inline Mirrlees engines, which were both quiet and reliable. Lifting a half-ton cylinder head on *Tiger's* V engines could be a real struggle, but doing the same job on *Cambridge Ferry* was a piece of cake. Another bonus was that we never carried any passengers, apart from a few lorry drivers, if we happened to be carrying HGVs instead of trains, although this didn't happen very often. This made ours a very laid-back sort of existence, and the management seemed happy to just let us get on with it; at Dunkirk if the weather was fine, some of the crew would even go fishing off the stern while we waited for our load.

Just to break the monotony, there were a couple of incidents I can recall that rather spiced things up. The first occasion was at Dover, where the trains were loaded in the old train ferry dock. This had an ingenious hydraulic gate to it, creating a lock. Once the ship was inside the lock, the gate was raised and the water level adjusted by pumps until the ship was at the right height for the trains to move on and off her. On this particular occasion we had a new engineer with us who was learning the job. He already had a chief's ticket and had told us he was studying for a degree as well, which he hoped would lead him somewhere better than simply going back and forth across the Channel on an old rust bucket like *Cambridge Ferry*. He had been with us for the best part of a week and I was in my cabin having a beer with Rob Caldwell, the fourth engineer, when the mate stuck his head around the door to say they would be ready to go in about 15 minutes and would we please start up. The new guy chimed in to say he would go down and do it while we finished our drinks. It was a very simple procedure which he had witnessed several times before and he had a chief's ticket after all, so I said, 'Yes, help yourself.'

After a few minutes we heard the familiar rumble as the engines fired up – but then a commotion started outside on deck, with shouts and running feet, followed by a big thump. It transpired that our new engineer had got the engines going without first starting the controllable-pitch propeller pumps. That omission meant that the prop blades would go to full ahead pitch, so the ship had lunged forward, snapping the mooring ropes, until she was stopped by the dock gate, by which time someone had got to the bridge and pressed the engine stop buttons there. Fortunately – and amazingly – neither the gate nor the ship had suffered any damage other than to the paintwork, so after Rob and I had gone down and pressed all the required buttons, we restarted the engines and the ship left. Surprisingly, perhaps, neither the chief nor the captain took us to task over this (I was sailing as second that week) but at the end of the shift the trainee departed and we never saw him again.

The second incident was at Dunkirk, where there was a conventional link-span connecting ship and shore, so the trains were loaded in the same way as the cars and HGVs on my previous ships. The procedure was for two trains to be loaded at once, one on each side, to keep the ship on an even keel. We did have trimming pumps that could transfer ballast from one side to the other if it turned out that the trains were of unequal weights, but mostly it needed very little adjustment after loading. The drivers on the shunting engines that were pushing the trains across the link-span could not see what was happening on board the ship at the far end and relied on signals given to them by the loading officer over their walkie-talkie radios. On this occasion the two trains were coming down together as usual – but then for some reason one of them stopped moving while the other carried on. By the time the second train had also been stopped, at least three more wagons had arrived on the starboard side than the port and the ship had heeled over so much that one wagon was actually suspended a couple of feet in the air above the link-span, held there by the buckeye couplings at each end. No attempt was made to move the offending train again until we had trimmed the ship upright, but as we did so and the hanging wagon descended back onto the link-span, one end of it missed the tracks. The French railway workers must have seen this sort of incident happen before, because it only took them half an hour or so to produce the necessary jacks and re-railing equipment to get the wagon back on the track. With this done, we could finish loading the train – which had to be done very slowly and carefully, one wagon at a time, as we had to adjust the trim for each one.

Today, Dover's Western Docks have been turned into a cruise terminal, and big ships tie up along the Admiralty Pier. The very imposing Dover Harbour railway station has survived, although there is not a trace of the huge network of tracks that once led to and from it; today it serves as the terminal building for the cruise ships. The train ferry dock has been filled in, and the old Granville and Wellington docks, together with the tidal basins, have been turned into a marina. Down at the Eastern Docks the submarine pens have gone and there is now a big terminal for freight ships. The last time I visited Dover I hardly recognised a thing.

After this little holiday, I was sent in early 1987 to a freight ship called *Seafreight Freeway*, also running between Dover and Dunkirk. She was altogether a more serious proposition, and just for a change the company had bought her from the Greeks instead of the other way around. Most things in the control room had been relabelled in English, but out in the engine room all the brass plates on the valve hand wheels were still in Greek, so for the first week or two we were struggling to find what did what and went where. I hated her from pretty well the word go: down below, the engines were filthy and dripping oil everywhere, while she also had ongoing exhaust leak problems adding soot to the mix. The engine room had very little headroom and was horribly cluttered – I didn't think it would be very easy to get out in an emergency. The engines themselves were a pair of V12 MWMs that were very noisy, although they seemed to be quite reliable and we certainly weren't having to change a pair of exhaust valves every other week as we had on *Tiger* – which was just as well, considering how little space there was to do the job.

I stuck it out for a few months but then rumours began to circulate that the company was going to streamline its operations, and redundancy loomed once again. Knowing this would be done on a last in and first out basis, I also knew that I would be one of the first to get the chop and that this time there wouldn't be a pot of money to go with it, so I started looking around for another job. My old friend Rob Caldwell also left around this time and had landed a job on the Antarctic survey ships, where he stayed for many years.

In May 1988 *Seafreight Freeway* suffered a major engine room fire, when the bolts holding down an oil filter cover had sheared and allowed oil to spray over the hot main engine exhaust manifold. One engineer was killed in the blaze and another badly burnt. I still count myself fortunate for having left the company when I did.

16 THE BOVRIL BOATS

As soon as I had decided to leave Sealink before I was pushed, I started scanning the situations vacant page of Lloyd's List, and after a few weeks found an advert by Thames Water, who wanted an engineer for its sludge carriers. These had long been nicknamed Bovril Boats, which no doubt was an allusion to the perceived colour of their cargoes. The salary was not great – probably about the same as a third engineer with Sealink – but it looked like it might be a steady job, which seemed to be a better bet than chasing a dwindling number of vacancies in Dover. I duly sent in my CV and before long was invited to an interview at the fleet headquarters at Crossness, near Abbey Wood in south-east London. I had an interview with the marine operations manager, who informed me that the company operated four vessels and worked a live aboard, week on, week off shift system with four weeks' annual leave. He asked me some interesting questions to test my knowledge of medium-speed diesel engines, and then said that they had a few more candidates to interview and would let me know in due course. This was the sort of thing that I knew they would say to someone who wasn't going to get the job, so I was quite surprised when less than a week later a letter came through the post, offering me the position. It turned out that there had been 101 applicants in total, so I felt rather flattered that I was the chosen one!

It is a popular misconception that the Bovril Boats took raw sewage and dumped it out at sea. This is totally false, as what we actually carried was the treated effluent left over after the incoming sewage had passed through the works. Sewage treatment is quite a complex process, involving (amongst other things), coarse filtration where larger solids such as lumps of wood and plastic are removed, maceration (chopping up), aeration, filtration and bacterial digestion. What goes in as raw sewage comes out as (a) clean water, which can be returned straight to the river, and (b) the sludge left over after the bacterial digestion. The latter is what we carried out to sea and dumped. A by-product of the digestion process is methane gas, which was burnt in the power house at Crossness and produced most of the electrical power required to run the plant. When I joined my first ship and we started loading, I was happy to find that our cargo did not smell like real sh*t, although the works itself could produce some pretty evil odours that wafted over us if the wind was blowing the wrong way!

Thames Water operated four ships. MVs *Newham*, *Bexley* and *Hounslow* were all the same and looked like old-style mini-tankers with the bridge amidships. They were each powered by a pair of Ruston 6-cylinder engines and were about 2,400 dwt. Lastly there was MV *Thames*, which was a bit bigger and had an all-aft accommodation block. She had a single Mirrlees engine but also had a bow thruster that the other ships lacked. My first vessel was *Hounslow* and for my first week I was sharing the job with Barry Dixon, her usual chief. What surprised me most about the vessels was how smart they were inside: the officers' mess was very well fitted out with easy chairs all around the bulkheads, a TV set in one corner and a dining table that would seat six. My cabin, although quite small, had everything I needed including a writing desk, chair and a comfortable bunk, all in mahogany with brass fittings – it was a cosy little den that reminded me a bit of *Post Runner*.

The crew of *Hounslow* consisted of the captain, chief and second officers, chief (and only) engineer, four seamen, two engine ratings, a cook and a steward, making 12 in total, which was very generous manning. In the officers' mess we were served at the table in exactly the same manner as on a deep sea ship, and with the opportunity for fresh food to be brought in every day except Sunday, the meals on board were always excellent.

The sludge fleet operations were determined by the state of the tides: we would always depart about two hours after high water and proceed down to the Barrow Deep in the Thames Estuary, where we would dump our loads, and then return again on the flood tide, the whole trip taking about nine hours. Because the time between high tides is almost half an hour more than twelve hours, it meant that some of our working weeks would consist of 13 round trips and others 14. When we changed crews at the end of each week, the joining time was always at high water, so it could be any time of the day. In case we hadn't been able to work it out for ourselves, the office would phone us the day before to let us know when and where to join the ship, as the fleet operated from the sewage works at Beckton, on the north side of the river, as well as Crossness.

When I had my first look at the engine room, I was pleasantly surprised at how clean and tidy it was, considering the ship was already nearly 20 years old. One thing that had helped to keep it that way was that we were burning nice clean gas oil as fuel, so there were none of the dark brown stains around the cylinder heads and down the sides of the crankcase that usually characterise an engine burning heavy fuel. Apart from the two Ruston main engines, we had three quite small generators on a platform raised one level above the engine room floor plates, any one of which could have handled the total electrical load of the ship, although we always had two running when on stand-by. I seem to recall that the engines did not drive CPP propellers but were fitted with a reversing gearbox and pneumatically operated clutches instead, so the propellers could be driven either way. Once the engines had been started and control handed over to the bridge, the engineer on watch had very little to do. In fact, as Barry explained to me, the engineer was only required to be on duty in the engine room from start-up at the beginning of the trip until we had passed Tilbury, and then again on the return journey from Tilbury upstream to the berth; this would only take about an hour and a quarter each way, or a bit more if we were running to Beckton. Apart from that, and if there was no maintenance work to do, we might just go up to the bridge and watch the river go by or even sit in the mess and read the paper; it was an absolute doddle of a job. When the engineer was not in attendance the two ratings shared the rest of the passage between them, and if it was a night-time trip, would wake up the engineer when the vessel was approaching Tilbury.

For my first week with Barry, apart from attending all the stand-bys, I spent a couple of days exploring the ship and learning all the valves and pipework, which was a lot simpler than on any other vessel I had seen. The sludge-dumping arrangement was interesting, in that it was all done by gravity. The tanks themselves did not extend to the full depth of the ship and you could in fact walk around in the spaces beneath them. When the ship was fully loaded the tops of the tanks were a few feet above sea level, so when the dump valves at the bottom were opened the sludge would start to run out, and at the same time the ship would rise in the water thereby keeping the level in the tanks above sea level, and the process would continue until the tanks were empty. By this time the ship would have risen so far that she would be too unstable for a safe return journey (not much GM!), so we would close the dump valves and take on seawater ballast for the trip back to Crossness or Beckton, where it would be pumped out again. The dump valves were huge – about 3 feet in diameter if memory serves – but thankfully they were power-operated and seldom needed any attention.

On Saturday mornings we tested the lifeboat engines and ran the emergency diesel fire pump, which was situated up for'ard – Barry sending me forth 'to amuse myself' as he put it. *Hounslow* and her sister ships actually each had a proper boat, not the RIB I had expected. They had a small jib crane to swing them out over the side and lower them away instead of the conventional davits. Starting the engines was done with a handle and you had to open a decompression lever first, then crank them up as fast as you could go and then drop the lever – basic, but foolproof. The fire pump was started in the same way, and in total this testing occupied me for about three-quarters of an hour.

Barry kept the engine room in pretty good order so there was not a lot to do in between trips, but during that first week I spotted a leaking gland on the domestic fresh water pump and re-packed it, and also re-jointed one of the crankcase doors, which was weeping a little oil – all normal routine stuff. Other than that and to help pass the time, I got out the Brasso and polished up all the usual gauges and handles around the engines.

On the last day of my introductory week, Barry called me in to the engineer's office and showed me all the paperwork for the week that needed to be done: tallying up engine and generator hours, stores and spare part orders, bunker requirements and the like. The last item on the agenda was the overtime sheets, and I assumed this must be for the crew. Much to my surprise, however, Barry said it was *our* overtime that needed filling in. Having always been salaried and never getting a penny in overtime with any of my previous companies (despite working quite a few 100-hour weeks for Esso), I enquired what I might have done to deserve any. 'Well, you ran the lifeboats and tested the firepump?' to which I nodded, and he said 'Well that's two hours' overtime – and then you packed the gland on that water pump and fixed that leaking crankcase door?' another nod, and another two hours for each job. 'And then

I see you've been busy with the Brasso, well that's another two hours, so you get eight for the week.' I noticed that he had given himself ten hours, two of which were for 'instructing the new engineer!' I protested than none of my jobs had actually taken two hours, to which he replied that two hours was the minimum payment and that apart from watching it all go up and down and round and round, any other work was overtime. It was a very generous hourly rate, and if it was the same every working week would pretty well take my pay up to what I had been getting in Sealink – I was amazed.

The deck officers did not miss out, either: they had brass to polish on the bridge, boats to swing out, charts to update and the like – it was all extra money for simply doing what any officer worth his salt would consider to be his normal duties. The really big money, however, was what you got if you were ever called back on your rest week to cover sickness. This would be four hours for simply turning up (and you got this even if it was just an extra day tacked onto your normal week, which would not mean any extra travel time), then you would get six hours for each tide you worked, with a minimum of two tides, so an extra day's work gave you 16 hours' pay at the overtime rate, which was very generous indeed.

All this extra money was the result of long-term agreements between the unions and London County Council, which had operated the fleet until the privatisation of the water industry, when Thames Water took it over. I think the union involved was the Transport and General Workers rather than the National Union of Seamen, and the TGWU had well and truly screwed the employers for everything it could get. Eventually, this led to Thames Water divesting itself of the burden and passing it on to Crescent Shipping, which tore up all the agreements and restored some semblance of normality on the basis that if anyone didn't like it they could take the redundancy money and leave. As this event was several years in the future, I simply reaped the benefits at the time, like everyone else.

At one time the steel handrails in the engine room had all been burnished and then coated with lacquer to stop them rusting again, but over the years the lacquer must have absorbed moisture as the steel beneath had darkened in colour and even had a few spots of rust showing through where the coating had been rubbed off. I thought that as I was going to get paid for it, I would try and restore them to their former glory, and set to with a roll of emery tape to get rid of all the varnish first and then polish them up. The stanchions were about 6 feet apart and it was about an hour's work to do a single section properly. I pressed on, though, and after about a month had done the whole lot and given the stanchions a fresh coat of black paint as well. I have to say that when entering the engine room after that, it all looked fantastic, and even Barry was pleased, after initially being rather dubious that I would ever get it finished.

Although any of the engineers could be expected to relieve on any of the ships when required, we mostly stuck to one regular ship and I actually spent at least two-thirds of my time on *Hounslow*. Of the others, the sister ships *Bexley* and *Newham* were of course where I usually went relieving. *Thames* was the odd one out, and I found her crew had a rather superior attitude to those of us on the smaller ships; in fact I only ever sailed on her a couple of times. I have to say though, that her engine room was the smartest in the fleet and every bit of brass and copper pipe was gleaming, while the single Mirrlees engine did not have a weep of oil anywhere.

Three engineers per ship was rather more than necessary to simply cover the working weeks, rest weeks and leave, so every so often I would find myself doubled up with Barry on *Hounslow*, which made what was already an easy job even easier – sometimes we would even argue about who was going to do any extra work that came up, as it was better to be doing something constructive rather than simply cleaning and polishing an already clean engine room or watching TV in the mess.

Although we were supposed to make 13 or 14 runs per week, depending on the tides, it very rarely happened that we actually did this many and we would instead spend the night on the berth at Crossness or Beckton. There were many reasons for missing trips: quite often there just wasn't enough sludge to warrant making one, and the other usual excuse was bad weather (usually fog) in the estuary. It was the captain's decision whether or not to sail, and I think that in Crossness especially, a trip might be cancelled simply so we could walk up the road to the pub for the evening, or so it seemed to me. Management took a very laid-back attitude toward this, and providing the sludge in the holding tanks was kept down to the required levels they left us to get on with it.

One time when a trip should definitely have been cancelled but wasn't happened on the night of 15/16 October 1987, when weather forecaster Michael Fish had famously predicted that there wouldn't be a hurricane. We had sailed in the evening as usual and got down to the spoil grounds before the full force of the wind hit us. I was actually in my bunk, happily in the land of nod, while this was going on and would have probably slept through it had not a seaman put his head round the door and said the captain would like me on stand-by in the engine room as he was concerned that he was having to give the engines a real thrashing. Down below it was noticeably warmer than usual and the exhaust temperatures were well up, although not enough for me to be overly concerned, so I phoned the bridge and reported all was well. At first, the captain opted to remain in the estuary, where he had more room to manoeuvre, but this left us at the mercy of some very steep and nasty seas – it was the one and only time on those ships that I ever saw waves breaking over the deck.

After an hour or two of this, the wind appeared to drop – probably as the eye of the storm passed over us – and the captain decided to make a run for home. Normally, even in the dead of night, one could easily navigate visually up the river past Tilbury, as the lights of London were all you needed to see by, but on this occasion London was blacked out because of the power cuts, and the captain had to find his way to the berth using the radar. All sorts of debris was coming down the river, including a couple of lighters that had broken free from their moorings, so we were lucky to get back to Beckton without hitting anything. Our troubles were still not over, though, as the wind had picked up again and the seamen were unable to get a rope ashore – every time they threw the heaving line to the men waiting on the jetty, it got blown back on deck. After about 20 minutes of thrashing around trying to keep the ship in position, the captain decided it would be safer to simply turn round and go back out to the estuary until things had calmed down, so that it is what we did, eventually getting back to Beckton at around midday on the 16th. Later that day we learnt that among several other casualties, the dear old *Hengist* had not been so lucky and had finished up on the beach outside Folkestone Harbour. When I phoned Astrid to see how things were at home, it was to learn that we had lost half the ridge tiles from the roof, our two greenhouses had been flattened – and that she had slept through the whole thing!

Working out of Crossness one could not fail to be aware of the giant Victorian engine house that stood to one side of the treatment works. This had been disused since 1953, when the original steam engines were replaced by electric pumps. When Sir Joseph Bazalgette had created the London sewage system back in the 1860s–70s, all the tunnels naturally had to run downhill and by the time everything arrived at Crossness it was some 30 feet below river level even at low water. To get over this problem, four enormous beam engines were installed to pump the sewage up again, so it could be discharged into the river on the ebb tide, hence the engine house. This removed the problem of raw sewage flowing into the river in Central London, much to the relief of the members of Parliament who had had to breathe the noxious odours from the river during the Great Stink of 1858 – but all it had really done was shift the problem further downstream. By 1891 sedimentation tanks had been added at Crossness, and from then on, only the water was pumped into the river while steam boats carried the resultant sludge out to sea. This was the start of the sludge fleet.

In 1987, when I started working there, the Crossness beam engines had simply been abandoned, as it was considered too uneconomical to break them up for scrap, although all the copper and brass had been removed. I was determined to get inside the engine house and have a look at them, even though the place was locked up and I couldn't find anyone who knew anything about it, let alone had the keys. Eventually I found a broken window and crawled inside, as no doubt several generations of youth had done before me, judging by the coke cans and fag packets I found littering the inside. As my eyes became accustomed to the gloom and I started to look around, I could barely believe what I was seeing – it was truly a cathedral of steam. The engines themselves were enormous, and you had to climb several very ornate iron staircases to reach the beams at the top, which Wikipedia tells me weigh 42 tons each. Everything was covered in 35 years' worth of dirt, dust and pigeon droppings, but even so there were traces of its former glory everywhere you looked: all the main support columns and the ironwork surrounding the stairs and platforms were ornately cast and had at one time been beautifully painted – there were even traces of gilding gleaming faintly through the grime. To say that I was astounded would be an understatement.

The people who had designed and built the place had succeeded in creating an absolutely wonderful piece of engineering which was both functional and very beautiful, yet in 1953 the doors had been closed and it had all been left to rot. On a construction of this scale these days, the bean counters would have put the red pen through all that unnecessary decoration and built something that did the job but wouldn't be worth a second glance – it would be, after all, just part of a sewage works, so who would want to see it anyway? Fortunately, around this time, a group of enthusiasts got together to start on the long road to restoration. Finally, after thousands of hours of unpaid hard labour by quite a small band of volunteers, one of the engines was returned to steam in 2003. Only those who, like me, saw the place before they started the project can have any idea of the massive task they had faced. Now there are regular open days, and for anyone with even the faintest interest in steam – or even those who haven't – it is a fantastic place to go and visit.

After two and a half years of going up and down the Thames earning pots of money for not doing a great deal, I heard rumours that things were about to change. In the first place, the method of sludge disposal itself was under scrutiny. Greenpeace among others was campaigning against the disposal of sludge at sea, and its aerial films of our vessels leaving their black trails at the Barrow Deep were compelling evidence to the public, who did not appreciate that what we were dumping was in fact little more than dirty water (actually it was 96 per cent water and 4 per cent solids) and certainly not raw sewage. It could not have been doing very much harm to the water quality, as even after a hundred years of sludge dumping, the area around the spoil grounds was a still a favoured spot for the Southend and Clacton fishermen. The alternative proposal was that a new plant be built, containing huge centrifuges to remove most of the water from the sludge, which would then be burnt in an incinerator. Some of the waste heat from the process would be recovered and used to generate electricity, but in my view all we would be doing would be to exchange doubtful sea pollution for definite air pollution – an opinion that was shared by Friends of the Earth.

The second point was that Thames Water, being a private company and basically profit-orientated, must have been wondering how it could rid itself of this very expensive operation. I think this was definitely an occasion where the crews of the sludge fleet killed off the goose that laid the golden egg. What the officers were getting with their 8 or 10 hours' overtime a week for brass polishing and the like was one thing – but it was the ratings who were the real culprits. It seemed to us that they had worked out an unofficial roster where a few of them would go sick for a couple of days during their working week but would be available to be called back on the rest week ,when the next lot went off sick and so on; this was like winning the pools. Weak management was of course to blame, and if the managers had shown some backbone and stamped on the practice a lot earlier, then we might have continued running quite happily until 1999 when the incineration plants were fully operational and sludge dumping at sea finished for good.

Thames Water eventually decided to sell the business off to Crescent Shipping, which was renowned for being a very hard-nosed operator, and before long it had set out its plans for running the fleet. Firstly, the roster was going to be changed to two weeks on and one week off, which would double our working hours for the same pay. Secondly, there would be no overtime for the officers; we would be salaried and that was that. Naturally, Crescent didn't expect us to simply take these changes on the chin and to avoid industrial action, Thames Water was forced to offer very attractive redundancy terms in an attempt to bribe us all out of our jobs, the idea being that it would get rid of as many of the original crews as possible and Crescent Shipping would start afresh with new faces, who would be happy to accept the revised conditions. For anyone who had been in the company ten years or more (which was the majority), the payouts were fabulous, and nearly everyone took the money. For myself, even though I had only been there three years, they were going to pay me almost as much as P&O had done in 1985, so I decided to go too. The writing was on the wall for the job, anyway, and I surmised that whenever it did finally come to an end Crescent Shipping would not be paying out anywhere near so much redundancy money as Thames Water had offered.

So that was the end of my short but happy time with the Bovril Boats, and my discharge book tells me that I finally signed off on 31 January 1990. I am not sure what became of *Hounslow*, although she was definitely still afloat and out in the Middle East in 2008 – I bet, though, that my lovely burnished handrails were not quite so smart by then.

17 FERRIES AGAIN, AND SEACATS

After leaving the sludge fleet I couldn't find any seagoing work straight away, so decided on a change of tack and applied for a job as an engineering instructor at the National Sea Training College in Gravesend. Despite the fact that I had no teaching qualifications, the principal of the college and the head of engineering were both prepared to give me a trial, so I started after the spring half-term holiday in 1990.

The college gave basic training to young school leavers who wanted a career at sea, either on deck or in the engine room. My job was to give very basic engineering instruction to these young people, and also to run the firefighting and confined spaces safety courses. The classroom work I detested pretty well right from the start, as the students had all left school early so they could go off to sea, but had immediately found themselves back in a classroom again. Unsurprisingly, for the most part I found it very difficult to instil them with any enthusiasm for the course. On one occasion I was talking about basic electrical knowledge and was trying to explain the difference between primary cells (throw-away batteries) and secondary cells (rechargeable) and asked the students to give an example of where the former might be used. A sea of blank faces looked back at me until one bright spark piped up with 'I fink that's wot my sister uses in her vibrator!' which caused the class to fall about laughing, although he was actually right (back then the only secondary cells most people encountered would have been car batteries). Luckily the college had quite an extensive library of films showing various aspects of life at sea, including engine room work, as I found that the most enjoyable, hence fruitful, lessons I gave were the ones when we started off with a film followed by a question-and-answer session.

As well as all the spotty-faced teenagers there was a small group of mature students, mostly engine room ratings, who wanted to learn enough to pass a basic engineering exam, which would eventually allow them to become engineer officers. They were much more amenable to instruction, and I felt that at least with them I was doing something useful. Like Mr Burrage at Poplar Tech, I would throw in the odd anecdote about my experiences at sea to liven things up during the more boring topics.

Running the firefighting course was quite good fun, although there was no mock-up engine room in which to terrify the students. Instead there were steel trays about

8 feet by 6, surrounded on three sides by brick walls, which made an impressive blaze when flooded with gas oil and ignited. We used these for hose practice and it was interesting to see how some students who rather fancied themselves as tough guys took no notice of my instruction to brace themselves when turning on the hose – that is, until the force of the water shoved them backwards, in one case dumping the operator on his behind, much to the amusement of his fellows. Another thing that never failed to grab their attention was the chip pan fire, which I extinguished with a fire blanket in exactly the same way as I had seen on my own course in 1974.

The confined spaces course consisted of a couple of days of classroom work, talking about the dangers of oxygen deficiency and the like, together with instruction in the use of breathing apparatus (BA). Fortunately, this course was only given to adults who had already been to sea, so I was spared my usual class of uninterested teens. For the practical test we used an old 40-foot shipping container with some internal partitions welded into it and a life-size dummy concealed inside. The students would enter through a trapdoor on the top at one end, climb down a ladder to the bottom and then have to zig-zag their way in the pitch dark through to the exit at the far end, collecting the dummy on the way. The technique was to use the back of your hand to feel around all the corners; either hand would do, but if you changed hands halfway you would find yourself back at the start. The students all went through the container once without BA, then with it. For the latter exercise, I would start a smoke generator so that when they opened the trap door to enter the box, dense smoke would billow out. The students did this exercise two at a time, with me following behind with a torch to use in case they ran into difficulties and had to be led out in a hurry. When anyone wears BA gear for the first time they nearly always tend to hyperventilate, and the majority of the students had used up most of their air by the time they exited the box, while in my case, having done it so often, I could make my air last for two trips with ample reserve. One pair had been so nervous about wearing the BA and descending into the smoke that they got through okay but forgot the dummy.

At the end of my first term, the head of engineering had a word with me to say that the college had been pleased with the way I had been getting on and was giving me a pay rise. He also wanted me to devise a course for basic fitting and machining work in the college workshops. These were well equipped with benches and machine tools, but they had never been used, as until I had come along the college had never had an instructor with the relevant skills. This occupied me for a couple of months, and I even got as far as cutting up material for the first set of exercises – but the truth was that I had already decided that teaching was not for me and that I didn't want to spend the rest of my working life there. In addition, it took me well over an hour each way to commute from Stelling Minnis, and the college itself was a rather depressing 1960s building in the middle of nowhere with marshes on three sides and the Thames on the other. The canteen in particular reminded me of Poplar Tech. and the food was equally grim, so I decided to start looking for a job back on the ferries again.

After a month or so I heard that Sealink had a few vacancies, and this time I was offered a second engineer's job straight away. This was rather better than I was expecting, and the salary was at least 50 per cent more than the college was paying me, so I accepted without hesitation. When I handed in my notice to the head of engineering, he seemed genuinely sorry that I was leaving, but thanked me for my efforts and wished me well.

In 1991, Sealink British Ferries was sold to Stena Line but the name lived on; the new company called itself Sealink Stena Line. For my first few months there I was a bit of a dogsbody and worked on several ships, including the H boats again, which I was quite happy about. If you look at the list of my ships in Appendix 2, you will see a number of vessels listed that haven't had a mention in the text, probably because I can remember nothing of interest that happened on them. Among them were *St Magnus*, which had come from Scotland in 1979 to provide a bit of spare capacity on the Dover–Boulogne route when *Tiger* or *Lion* were in refit. The only thing I can remember about her was the smell of cattle on the vehicle deck, as apparently she had regularly been used for carrying livestock. Then there were *Vortigern* and *St Christopher*, which both belonged to Sealink and on which I think I spent just a couple of weeks each on relieving duties.

Finally in this list of also-rans was *Stena Fantasia*, on which I spent less than a week but which I certainly do remember. I had been sent to her to learn the job before taking over as second, and was very pleased to come across Steve Rooke from *Flinders Bay* again who was serving as relief chief engineer. After I have spent just a few days on board, *Fantasia* was caught in gale force winds that drove her diagonally into a corner of Dover Harbour, where she was relentlessly battered against the quay. She was a huge ship compared to an H boat and simply did not have enough power to push herself off against the wind, even with assistance from the harbour tug, so we just had to watch as the starboard quarter of the ship slowly got smashed in against the concrete. The hydraulic pumps and controls for the stern door were situated right next to this area, and after about half an hour the side of the ship had been squashed in so much that all the hydraulic pipes were fractured and oil was escaping everywhere. Eventually the wind dropped and we were finally able to tie up on our berth and check the damage. It was plain that we were never going to be able to open the stern door, so all the vehicles had to remain on board until we could get the ship turned around and offload them through the bows instead. Finally, she was sent off somewhere for repairs and I never saw her again.

Around May 1991 I was asked if I would go out to Denmark to spend a week as part of a group of officers learning how to operate a ship called *Peder Pars*, which Stena had just bought and was going to operate out of Dover. Myself, two other engineers and two deck officers were sent to Kalundborg, on Zealand, where we were boarded in a small but pleasant guest house. After breakfast every day, we would troop down to the ship and stay on her for one return trip to Aarhus, on Jutland. This was just 40 nm each way, so it was all very laid back, and in the evenings we would go into the town and find a bar or restaurant, so it was quite a pleasant sojourn. We had been provided with a car to get us to and from the docks, and one day when the sailing was cancelled for some reason, we decided to drive into Copenhagen, where we did some sightseeing.

After Kalundborg, it was some weeks before we next saw the ship, as she had to go off to refit, where she was painted in Sealink Stena colours and renamed MV *Stena*

Invicta. She also needed modifications to the bow and stern doors in order to fit the berths in Dover. The modifications at the bows were particularly ugly; they consisted of pair of vertical columns welded on at each side, which supported a horizontal girder arrangement upon which the bow ramp could be lowered. Whenever there was anything much of a sea running, this girder would be the first thing to hit it and the resulting sheets of spray could be quite spectacular.

Stena Invicta was very much bigger than anything else I had sailed on out of Dover, and could carry around 400 cars and no less than 1,750 passengers. Instead of the usual medium-speed V engines, she was powered by a pair of 8-cylinder B&W two-strokes of 8,480 hp each, driving CPP propellers. After being started, they were run up to their full speed straight away because they each had a shaft generator which between them provided all the electrical power required at sea; they were never idled. This meant that the normal diesel generators could be shut down when on passage, making her a very quiet ship. When in port or under stand-by conditions, the main electrical load was provided by two out of the three diesel generators, while the shaft generators were used for supplying the two bow thrusters and the stern thruster.

At the end of each trip, changing over to harbour conditions was quite a performance: first off, two of the diesel generators had to be fired up and manually synchronised with the shaft generators. When they had been connected the load was transferred to them by adjusting the governor controls until they were taking the lot, at which point the shaft generators could be disconnected and then reconnected to supply the thrusters instead. At least, this was what the other crews did – but there was a simpler way. When I had been out in Denmark I had watched the Danish engineers do this several times and they simply flipped a single switch from the 'at sea' position to the 'harbour' condition and the whole lot was done automatically. For some reason everybody on *Invicta* seemed scared to death of doing this, but as I had by now been made chief engineer again, I decided to give it a go. When the bridge called Point Alpha (3 miles off the harbour entrance) my crew all stood back very apprehensively as if expecting an imminent blackout, but I was not to be deterred and flicked over the magic switch. There was a couple of seconds' delay and then we could hear the diesels firing up. A few seconds more and there was a succession of clonks from the switchboard as the various circuit breakers went in and out – and then it was done, having taken about a quarter of the usual time. After that, everyone used the switch.

Stena Invicta was a very good ship to work on, and the engine room was quite spacious compared with those on all my other ferries. She was also very reliable and I can only recall one incident which gave us some brief heart palpitations: we were in Dover at the time when the alarm for low air pressure went off, and a quick look at the gauge showed that it was going down fast. It was obviously a major leak which we should have been able to hear, but after a quick scoot round the machinery spaces we were still none the wiser. By this time the air pressure in the engine room was dangerously low, when the fourth engineer happened to be passing the diesel generators and luckily noticed that one of the gennies was slowly turning over, when it was supposed to have been shut down. The air start solenoid valve had opened for some reason, and it was a matter of seconds to shut the valve manually, after which we could all begin to breathe normally again. The engine could not have actually started as the fuel was shut off – but this had not stopped it from turning over and gobbling our precious air up.

I would happily have stayed on *Invicta* for many more years, but once again the company I was working for was looking to make economies. The first thing to go was the Folkestone–Boulogne service: the H boats were sent off to the Irish Sea routes, while *St Christopher* had already gone. I was now looking at my fourth redundancy in six years and was wondering, not for the first time, whether I would have been better off staying with Townsends in 1985. As it was, within a couple of weeks I landed a third engineer's job with P&O in Portsmouth, a bit of a comedown after being chief on *Invicta* – but beggars can't be choosers, as the saying goes.

The ship I was sent to was an ex-Townsend Thoresen vessel, *Pride of Winchester*, running a daily service to Cherbourg. She followed established Townsend practice: triple screws but only a single rudder, the idea being that as the middle engine would mostly be going ahead the resultant wash over the rudder would push the stern in whichever direction was required, even when going astern. (All the captains I spoke to reckoned that it would also mean that the wing engines would have to work harder when going astern and that the system on *Lion*, *Tiger* and *Panther*, with twin screws and twin rudders, worked every bit as well.) Power was provided by three Wartsila medium-speed diesels, the wing engines having eight inline cylinders and the centre engine nine.

We were working a rather odd service from Portsmouth to Cherbourg: the ship was actually fast enough to make two round trips a day but while I was there we only ever did one. We would sail around mid-morning and make a fast passage across, arriving around 2 pm, if memory serves. Then the ship would sit on the berth until the evening and would make a really slow overnight return to Portsmouth, arriving around breakfast time. For this return leg we would shut down the middle engine once under way, and stooge along at barely 10 knots, so it was a really peaceful trip. The roster system was a week on and a week off living on board, so it was basically the same as in Thames Water – but without any overtime for polishing the brass.

I had been put on the 8 to 12 watch, which suited me very well: the evening watch was spent getting ready for sea as we left Cherbourg and doing the first two hours of the crossing, following which I could get almost seven hours in my bunk if I wished. The morning watch would be much the same: get the engines ready for sea and then the first two hours of that trip, after which I would have a nice peaceful afternoon before relieving the 4 to 8 for dinner. We seldom worked field days, so it was not terribly demanding. On the other hand, it didn't seem that there was going to be any chance of promotion in the foreseeable future and I can't say I was terribly happy about having to clean centrifuges and other similar jobs which I had thought were well behind me. The bottom line, though, was that it paid the mortgage and I was still spending every other week at home, so it could have been a lot worse.

I had started the job in Portsmouth in November 1991 but by the following spring when the weather had started to warm up I would bring my bicycle down to the ship for the week. As the afternoons in France were usually free, I could go for rides in the surrounding countryside, on one occasion reaching Barfleur, a 40-mile round trip. There were masses of lanes to explore and it was quite easy to get lost, so usually I would not go so far. After one excursion I was returning to the ship and was going through the town centre of Cherbourg, when I was brought to a sudden stop at a set of traffic lights. I was using the old-fashioned toe clips and straps to keep my feet on the pedals and had forgotten to ease off the straps, so when the bike stopped moving, I was unable to get my foot out of the clip in time and fell over complete with bike, which gave the passers-by a good laugh at my expense.

Despite this job being an easy one and without having any immediate question marks hanging over the future of the route, it wasn't long before I started to get itchy feet and started looking for something else, preferably nearer to home. After a week or two, I saw that Hoverspeed was looking for engineers for its Seacats, which were working out of Dover and Folkestone to Calais and Boulogne respectively, so I applied and after a very brief interview was offered a job. The salary was excellent and I was going to be sailing as chief engineer again, so I accepted with alacrity and started in April 1992.

The Seacats had originated in Australia, where they had been designed by a company called Incat (International Catamarans) of Hobart, Tasmania. They were a quantum leap in size from any previous high-speed passenger-carrying catamaran and were 240 feet long with a beam of 85 feet and could carry about 80 cars and 450 passengers. Power was provided by four Ruston V16 four-stroke diesels of 5,000 hp each, running at 750 rpm and driving water jets. Because the engine room compartments in each hull were only 11 feet wide, the engines had to be angled slightly to enable the propeller shaft of the for'ard engine to pass its companion, immediately behind. The engines were pretty well self-contained units, with integral lubricating oil and cooling water pumps, and apart from a small starting air compressor there was little other machinery in the engine rooms. Electrical power was supplied by three small Caterpillar diesel generators at vehicle deck level, one on the port side and two on the starboard.

The craft was what was called a wave piercing design, and the two hulls had very fine pointed bows with minimal buoyancy, so that when they met a wave they would pierce through it rather than ride over it, reducing the stresses on the hull and giving a smoother ride – at least, that was the theory. The central part of the hull was normally clear of the water – in fact, when doing hull inspections we would use the craft's own small rescue boat and cruise around under it – but if bigger waves were encountered it had a heavily flared under-section and keel, looking a bit like the bows of a power boat when out of the water. This would provide extra buoyancy and thus lift the hull as the wave passed under it.

There was a bridge deck and beneath it two main decks: the lower of the two, the vehicle deck, was out of bounds to passengers during the crossings, while the main passenger deck had row upon row of airline-style seats, each with a seat belt (which turned out to be a very useful precaution). Apart from the seating, there was a duty-free shop and a small area at the stern for those passengers desiring a bit of fresh air. The bridge deck was very small and was reached by a staircase up from the passenger lounge. There was also a small area behind the bridge that was open to passengers and they could look through the windows and actually see what the bridge crew were doing. There was no galley, and any food served on board was brought aboard in insulated metal boxes, as on an aircraft.

My first Seacat was *Hoverspeed Great Britain*, which some may recall was once the holder of the Hales Trophy, otherwise known as the Blue Riband of the Atlantic, awarded for the fastest Atlantic crossing by a passenger ship. *Great Britain*'s award had come about because on her delivery voyage from Hobart to the UK she had gone via Panama and then to New York, where she had waited for good weather before making the dash. Leaving New York on 19 June 1990, she arrived at Falmouth on 23 June after completing the crossing at an average speed of 36.65 knots, beating the 38-year-old record of USS *United States* by about three hours. It has to be said, though, that compared with *United States* one could hardly call the *Hoverspeed Great Britain* a passenger ship at all: she had no cabins, for a start, and I imagine that the crew and passengers alike would have had to sleep on the deck on mattresses – not quite in the same league for comfort as some of the previous record holders, which apart from the *United States* included *Mauretania*, *Queen Mary* and *Normandie*.

The hovercraft's crew consisted of a master, chief officer, chief engineer, assistant engineer, a few seamen (still all male, even in the early 1990s) and a great many cabin crew. On the bridge there were three seats: the master sat in the middle with the throttles and steering, the chief officer was on his right and usually spent most of the time with his head in radar set, plotting a course that would avoid all the other traffic as we crossed the world's busiest shipping lane at speeds of up to 40 knots, the chief engineer was on the left at a console displaying all the engine parameters and controls, while the passengers could stand behind us with their noses against the windows watching the performance. It was all very different to what I had been used to.

At first I quite enjoyed whizzing across the Channel in the Blue Riband holder, and it was great to be able to see everything that was going on as we went in and out of harbour instead of being in the control room watching the engines getting thrashed and wondering how big the bump was going to be when we hit the quay. It was not long, though, before I started to see the deficiencies of this type of vessel – and there were many. You could start off with the crewing arrangement, where I was the only person on board with any technical training. The assistant engineers – at least, all the ones I came across – were not proper engineers at all; indeed, my regular guy had been a driver for the East Kent Bus Company only the previous year, although he said he also had some experience working as a garage mechanic. He knew how to top up the oil and water in the engines and generators, but he was not someone I could share a technical problem with.

Hoverspeed seemed to be trying to run its Seacat operation in the same way as an airline: the crew were only there to get the craft from A to B and were not expected to do any serious repairs, so Hoverspeed had thought it unnecessary to provide an engineers' workshop and there were precious few tools either. This meant that even quite a minor job could cause us big problems and I normally kept a few of my own tools, including a socket set, on board, while one of the other chiefs had brought in a bench vice that he had bolted into a recess off the vehicle deck. The lack of an onboard electrician was my biggest grumble, however, as in the event of a blackout (and we had several) there was no one to help me. The company view was that they had an on-call electrician in Dover 24/7 who could give advice over the VHF, and who would always be available to meet the vessel on the next arrival – assuming that is, that I could get it fixed sufficiently well to actually get it back to Dover.

Turning to the craft itself (we nearly always referred to them as such and certainly never as ships), we could start with the hulls, which were of all welded aluminium construction and were prone to cracking in the most heavily stressed areas, especially down aft in the jet rooms, which were very unpleasant places to be when at sea. A water jet is basically a propeller in a tube, and the jet rooms which contained them were very small compartments situated at the very stern. The noise and vibration there were the worst I have ever experienced – even with ear defenders it was deafening – and just to add a little extra excitement there were nearly always fine jets of water spraying through the sides of the hull at the aforementioned cracks. When you climbed out from a jet room onto the mooring deck above you got the best impression of the power and speed of the craft, with the thundering plumes of water exiting the jets rising up several feet above head height in the wake. To see this on a moonlit night, looking from one hull across to the other with the sea racing past between the two, was an unforgettable experience. To give you an idea of the power of these water jets, their output was not measured in gallons per minute but in tons per second – and that was for each jet.

Although there were four engines, only the inner two could be manoeuvred. There were no rudders, so directional control was obtained by angling the water jets up to 15 degrees to either side – 'vectored thrust' being the official terminology. For the craft to go astern, each water jet had a curved flap called a bucket, which could be pivoted down to interpose it into the stream of water from the jet, deflecting it forwards and thereby providing reverse thrust. The outer, or boost, engines as we called them, were there simply for extra speed and we would not normally start them up until we had left the berth and were heading for the harbour entrance. Naturally, if one of the steering engines failed, the craft became extremely difficult to handle and directional control could only be obtained by balancing the thrust between the two hulls, while if it was necessary to swing the vessel around when berthing, a tug would be required. To those of us operating the vessels, it seemed incredible that in order to save a few thousand pounds we had been left with this highly unsatisfactory arrangement. If one were to compute the cost of lost sailings and tug fees, not to mention the occasional damage repair cost, the extra money spent on making all four engines manoeuvrable would soon have been recouped.

Turning to the engine rooms, the first thing to strike me was that there were no lubricating oil centrifuges, and the engines relied on simple filters instead, just like a diesel car or truck. On the latter, one would change the oil and filter about every 12,000 miles, which at an average speed of 30 mph overall would be every 400 hours. We would rack that up on the Seacats in about six weeks, but we certainly didn't change the oil that often. We regularly changed the filters, but they were nowhere near fine enough to remove the carbon particles that would then find their way past the piston rings and into the sump. We would send oil samples off to Castrol every month for testing, and the results would come back telling us that the oil needed changing – but we seldom had enough to do that. The general opinion among the engineers was that we didn't think Hoverspeed had been paying the bills, so we didn't get the oil. I can't remember exactly how much each engine sump contained, but I think it took four or five 205-litre drums to do the job, so with four engines it was a very expensive business. If they had only installed a small centrifuge in each engine room it would have extended the life of the oil almost indefinitely and paid for itself in less than six months. Apart from *Esso Preston*, which of course had no sump oil, all my other ships had centrifuges, so oil changing was almost unheard of. Running habitually on filthy oil will eventually have consequences – but more on that subject later.

Another difference was the lack of a conventional bilge-pumping system. On the Seacats, instead of a pump in the engine room and a bilge main leading off to all the other compartments each with its own bilge well and suction valve, there was a separate submersible bilge pump in each compartment, which could be started and stopped by the engineer on the bridge when its respective bilge alarm was triggered. By the time I joined the craft, several of them were already defunct and had been replaced by the kind of submersible pump you might get from a hire shop to empty a small pond or a flooded cellar. The jet rooms, which were always filling with water from the many leaks, usually had one of these, and the discharge from the hose would simply empty out over the vehicle deck and run away through the scuppers. More than one motorist enquired of me where all the water was coming from and I would have to tell them we were washing out the space rather than say it was because we were leaking like a sieve.

Once we had commenced a crossing, and provided everything appeared to be okay at the console, I would make an inspection of all the machinery spaces and jet rooms. The only problem with doing this was that once I was in the engine room it was almost impossible to communicate with the bridge because of the noise – even with an aircraft-style plug-in head set, two 5,000 hp V16 diesels running at full chat in such a confined space usually overpowered any attempt to hear what was being said. Another problem was that the bridge was the only place where you had the full set of instruments and alarms to tell you what was happening on each side. If I was down in one of the engine rooms I would be able to monitor everything for that side, but if in the meantime something went wrong on the other side I wouldn't be able to check it out unless I went over there or back up to the bridge first, which was crazy. The master would be able to spot any red lights on my panel – but of course it was almost impossible for him to phone down and tell me what had shown up, because of the noise down there.

Hoverspeed, I think, considered itself to be a cut above the other ferry companies, and was very lax about traditional customs such as signing ships' articles. In fact, I had to make all my own entries in my discharge book to cover the period I worked for the company, although I got them countersigned by the captain, and my final entry has an official stamp. This lack of detail makes it almost impossible for me to tell, looking back now, exactly what happened on which vessel and when. Apart from *Hoverspeed*

Great Britain (Incat hull number 025), I also served on *Hoverspeed France* (Incat 026), *Hoverspeed Boulogne* (Incat 027) and *Seacat Scotland* (Incat 028), although the bulk of my time was split between 025 and 027. I have quoted the Incat numbers because the craft were all renamed many times during their lives, so if you want to know which went where and when you would be better off searching by hull number.

When *Great Britain* had crossed the Atlantic, it first went into service on the Portsmouth–Cherbourg route, where it quickly earned the nickname of the 'Vomit Comet', as the two-hour crossing in anything other than a flat calm would make the majority of the passengers sick. The problem was not excessive rolling or pitching but the sheer violence of the motion – likened to being on a bus driven fast through a slalom course over speed bumps. In the Dover Strait, where our crossing times were less than half that, it would still make a lot of the passengers ill at anything above force 5, and nearly all of them at force 7 – and we got a lot of that out of Dover during winter. One had to admire the cabin crews, who had about three-quarters of an hour to clean up all the mess and get the passenger lounge looking (and smelling!) nice and fresh for the next lot of victims, and they sometimes had to do this several times in a shift.

Apart from the sea sickness, we got quite a few minor injuries too, as despite being told to stay seated and use the seat belt (and the very nice Hoverspeed sick bag), quite a few passengers would try and stagger to the loo or outside to get some fresh air, and could easily be thrown off their feet if they weren't hanging on tightly. We had an accelerometer on the bridge which measured the vertical G forces and it would occasionally read 0.5 and over. This may not sound much compared to what a fighter pilot has to endure, but it means that a 12-stone man standing up would alternately be buckling at the knees when his weight suddenly became 18 stone and then be feeling light as a feather when it dropped to 6 – and this would be repeated every few seconds. On the vehicle deck it was fascinating to watch all the cars rising and falling on their suspension in unison, like a vehicular exercise class doing its knee-bends.

Apart from the cross-Channel routes, some of us were sent up to Stranraer on *Seacat Scotland* (028) for a few weeks to start running a service to Belfast. I was interested to see how the craft would get on when up against the big Atlantic swells when we rounded Land's End on the delivery trip, but much to my surprise I found that it handled these a lot more comfortably than the short, steep seas of the Dover Strait. Whilst up in Scotland we were billeted in what was probably the cheapest and meanest guesthouse Hoverspeed could find, so I was more than happy when my spell of duty came to an end and I could drive home, having taken my car up with me on the craft.

Working from Folkestone to Boulogne, there was another problem that happened periodically: fishing net getting sucked down the jet tubes and wrapped around the propeller. Coming out of Boulogne, the route is due north to start with, skirting the coast and very close to the fishing grounds. The nets were usually marked with small flags, which from the bridge high up on a big ferry were fairly easy to spot and avoid – but low down on a Seacat travelling at 35 knots or more, we could be over them before we had a chance to see them. When this happened, there would be a horrible vibration that could be felt throughout the craft, as the propeller would now be badly out of balance.

To find out which was the affected engine the captain would pull back each throttle in turn to see which one cured the vibration. If it was a steering engine we would probably keep it running at reduced speed, as it would be needed when we were docking; if it was a boost engine, though, we would shut it down. Next, we would contact a firm of divers based in Dover, who would meet the craft on arrival and clear it for us. This meant that the diver actually had to swim up the jet tube to reach the propeller, and I would go down to the engine room to shut the starting air valve to the engines and also remove the fuses from the control circuits to ensure that nobody could press the button by mistake and convert the diver to mincemeat. If it had been me doing the swimming, I would personally have gone below and checked these things for myself, but the usual guy we had (this happened three times during my brief spell in Hoverspeed) was very laid back and would simply take my word for it – a brave man indeed!

Going back to the dirty oil problem again, the test results coming back from Castrol gave us a whole lot of information on viscosity, additive levels and other stuff, but what we were most interested in was how much solid matter the oil contained. As mentioned previously, these solids were the very fine carbon particulates that were small enough to pass through the filters. When the concentration of these reached 3 per cent, we were supposed to do an oil change: I can remember one set of tests that came back where all four engines were over the limit and the worst had 7 per cent!

Running on black sludge like this was obviously not going to do the engines any good, and eventually we started to have problems. These would invariably manifest themselves any time a trip had been missed, as that would give the engines a chance to really cool down. The Ruston engines were quite finely engineered compared with a lot of the stone-crushers I had seen before, being less than half the size of the engines on *Tiger* and *Lion* but producing almost as much power and the clearances between the exhaust valve guides and spindles were very fine. If this clearance was further reduced by burnt on layers of carbon, when the engine was started from cold, the exhaust valve spindles would heat up and expand more rapidly than the guides, using up what little clearance there was left and cause them to seize up, which of course meant that the compression was lost on the affected cylinders. The governor meanwhile would apply more fuel to maintain the required rpm, which would heat up the other valves even more rapidly, causing more cylinders to die, and finally the engine would stop with half the exhaust valves stuck partly open.

After this it was pointless trying a restart, because there wasn't enough air to give it more than another couple of short attempts, which would not be enough to free the valves. The air receiver was very small, and the compressor that recharged it was not much bigger than the sort you might find in a garage for pumping up car tyres. This was yet another example of how poorly the craft compared to conventional vessels, where it was a requirement that there should be enough air for at least eight starts per engine.

For some reason (sod's law again maybe), it was nearly always a steering engine that suffered this type of failure, so the only option we had for sailing, providing the weather was okay, was for the captain to take us out on the other three, balancing the thrust between the two hulls to keep us pointing where we wanted to go. Once outside the harbour, we would head northwards along the coast towards Deal so we did not have crossing traffic to worry about, and would go as fast as we could. Meanwhile, down below I would open all the indicator cocks on the dead engine, which would release any remaining compression, and it would then start freewheeling, being turned by the wash over the propeller. After a few minutes with the offending valves being knocked down by the push rods and then brought back at least part of the way by the springs (the clattering sound from the valve gear while this was going on was enough to make any engineer wince), they would free off, and I would start to hear sharp hisses from the cocks indicating that the compression had been restored. At this point I would shut all the cocks, take manual control of the governor and very slowly give the engine some fuel, which would then start firing. I would maintain this very low fuel rate for several minutes more until I was sure that it wasn't going to die again, and then hand control back over to the bridge. We called this procedure bump-starting, and I had to perform it on at least three occasions.

On one evening, we were running into Calais on 027, the bow visor had been lifted and we were approaching the berth at no more than a fast walking pace. When we were about 30 yards off the vessel suddenly blacked out without any warning at all. Under these conditions, the engines would drop down to idle speed, but all steering and reversing control would be lost. There was nothing the captain could do apart from hitting the engine emergency stop buttons, which of course did not actually stop the craft, so we coasted in until we were stopped by the visor hitting the end of the jetty. The impact was sufficient to fracture both of the hydraulic rams that supported the visor, and as the craft rebounded the visor fell back down with a crash that shook the whole craft. By this time I had arrived at the switchboard, fired up a generator and got the power back again, so we were able to restart the engines and get the craft tied up. I tried all three generators in turn and could find no fault. As we were now unable to get the cars off through the bow, we had to turn the craft around and back them off the stern instead, which took quite a while.

By a stroke of luck, when the visor fell it had landed pretty much in the right place, so after a bit of bashing with the heaviest hammer I could find I got the locking pins in and we were safe to put to sea and return to Dover for repairs. The on-call electrician attended but could not find anything wrong, so I am quite sure the company thought it was somehow my fault – indeed, one of the managers came down from the office demanding to know what I had been 'playing at' and asking for a full report. I duly wrote something along the lines of 'Everything was quite normal and then we blacked out', which was probably not what they wanted to hear, but the captain backed me up and countersigned it, so that was that. Two days later, after repairs had been completed and the craft was back in service, the same thing happened again with another crew, but this time in mid-Channel where no harm was done. They too called in the electrician – a different one this time, who did find something wrong – I think it might have been a voltage regulator problem and after it had been changed we had no more repeats. Hoverspeed however, never had the courtesy to apologise to me for pointing the finger of suspicion.

Early in 1993 I was sent off with 025 to Newhaven for refit. I had written a list (actually it turned out to be little more than a wish list) of all the immediate repairs required, together with a few other things that I thought were necessary to make the craft more reliable and safer. Top of the wish list came a pair of centrifuges, but needless to say the red pen went through that; this was a decision that was to have expensive consequences within weeks of the craft going back into service. I also asked for the company to consider fitting the two boost engines with manoeuvring gear. This would have meant getting the craft into a drydock and would have cost a great deal more than the centrifuges – and this too was refused. Among the things we did get, though, were repairs to the leaking jet rooms. I had fondly imagined that this would mean trimming the ship down by the bows to raise the cracks above the waterline, then grooving them out and rewelding. What we got were aluminium patches pop-riveted on, with rubber sheet beneath to form the seal – I kid you not!

While we were there we had the Department of Transport surveyor down to check the craft out and issue a new passenger certificate, without which the vessel could not sail. This was a wonderful opportunity to explain to someone with real authority all its deficiencies, and I did my level best to convince him of the need for modifications, notably the fitting of manoeuvring gear to the boost engines – but without success. He only seemed to be interested in checking whether the life raft overhaul certificates were up to date and whether we had the right number of fire extinguishers; it seemed little more than a box-ticking exercise. Whatever the rights and wrongs of this – and I have my own private opinions about it – the surveyor signed the certificate and the craft returned to service in pretty much the same unsatisfactory condition as before.

It was only a few weeks before the next misfortune when one of the main engine turbochargers failed with bearing trouble. In view of our almost total lack of tools and facilities changing a blower was beyond the capabilities of me and my assistant, so a team of fitters was put aboard to do the job: it took them a whole day and meant at least four lost sailings. The reason for the failure was that the blower was running on the same filthy oil as the engine. Every other marine turbocharger I had seen had its own separate oil supply and could therefore use a dedicated turbine oil for the bearings. I would have thought it was obvious that a blower spinning around on ball races at 20,000+ rpm is not going to want the same sort of oil as a diesel running on plain bearings at 750, and the only reason I can see for the arrangement we had was that it must have been cheaper initially. As it was, I am sure the cost of the exchange blower, the lost revenue from the cancelled sailings, and the fitters' time, would have far exceeded the cost of a centrifuge: I have just gone online and found several of the size we would have needed for less than £8,000 each (second-hand, reconditioned) – and this is at 2020 prices.

Several more weeks passed before we had another incident – this time a lot more serious. We were on passage and everything appeared to be normal. I was sitting at my station next to the captain and monitoring my panel when I noticed a slight rev drop from one of the port engines. I just had time to say to the captain that I thought we might have a problem when the CCTV on that engine room showed a huge gout of flame and smoke erupting from the explosion doors on the side of the for'ard engine, swiftly followed by the fire alarm going off. I stopped both port engines as the CCTV was now blacked out by smoke and there was no way of knowing if there was still a fire, while the captain brought the other two back to idle and then stopped them too, as with no thrust at all on one side the craft would go in circles. Our passengers, meanwhile, looking in through the windows behind us, had also been able to see the CCTV and must have been wondering what was going to happen next.

We broadcast for the emergency party, which included me, the mate, one seaman and the assistant engineer to assemble on the vehicle deck. The assistant engineer didn't show up, and I assumed he must still be in the engine room, so I put on breathing apparatus and went looking for him. It was a most unpleasant experience having to grope around in the thick smoke while I carried out my search. After I had gone the full length of the engine room and circumnavigated both engines, I had found nothing and the whistle was now going off to let me know that my air was running out, so I climbed the ladder and escaped into the fresh air, red-faced and bathed in sweat. My assistant had by now arrived and said he had been in the loo and hadn't heard the call. I had the strong suspicion that he had been in a stock room with a stewardess with whom he had rather more than a professional relationship – what I called him would be quite unprintable.

There had, as it turned out, been no fire, and as the aft engine was okay we started up again and proceeded on three engines to Dover, where I could make a proper examination of the damaged one. What had happened was that another turbocharger bearing had failed, except that on this occasion it had completely self-destructed and some of the red-hot shrapnel had gone down the oil return pipe to the main engine sump, where it had ignited the oil mist therein and caused a crankcase explosion. All large marine diesels are fitted with spring-loaded explosion doors on the crankcase, which in the event of an incident like this will blow open and release the pressure, after which they are designed to spring shut again, to prevent more air getting back in and causing a secondary explosion or a fire. Fortunately, in our case they had done their job. On our arrival back in Dover the craft was once again out of service for a couple of days while a new turbocharger was procured and fitted.

What with bump-starting, the bow visor incident and now a crankcase explosion, I was starting to feel decidedly twitchy about Seacats. I had been sleeping badly and was always on edge waiting for the next crisis. I was particularly worried that in the event of a more serious accident I might find myself at a court of inquiry, where some highly paid barrister employed by Sea Containers would try to ensure that I, rather than the company, would get the blame for any disaster. It would be useless pointing out to the court that its attention had been drawn to the deficiencies of the craft on several occasions before – in this sort of case it always seems to be the senior officers on board at the time who get the blame, while the company would simply have had its knuckles rapped. I was beginning to think it was time for me to move on yet again.

Since leaving P&O in 1985, I had been to Sealink, then Thames Water, back to Sealink (Stena), P&O in Portsmouth and now Hoverspeed, so perhaps it was time for a complete change. I started scanning the situations vacant column in the *Daily Telegraph* instead of Lloyd's List. After a while I spotted a job with Cornhill Insurance, which wanted an engineer surveyor with pressure vessel experience, so I sent in my CV and awaited results. I was already older than the advertised age limit, but was interviewed anyway and got offered the job. The salary was roughly two-thirds what Hoverspeed had been paying me, but I got a company car as well and I would be working with steam again. Most of the inspection work would be on industrial boilers and pressure vessels, but with the occasional traction engine or railway locomotive thrown in to provide some real interest, so after some serious thought, I accepted the offer and gave Hoverspeed a month's notice.

I had some leave owing, which meant I didn't actually have to work many more shifts, and my last one was on 28 April 1993 on *Hoverspeed Great Britain* working from Dover to Calais. For the final arrival back in Dover it was a pretty well flat calm evening and the captain asked me if I wanted to take the controls and berth the craft, which was very decent of him. So I took position on the starboard side of the bridge and brought the vessel in, with the captain standing beside me giving instructions when required. He never actually had to snatch the controls out of my hand, so I must have done a reasonable job.

Having tied up and stopped the engines for the final time, I sat back in my chair and awaited the arrival of my relief. When he came up the stairs onto the bridge I was able to tell him, 'Just for a change and apart from having to bump-start the starboard inner last weekend, there's nothing wrong with the ruddy thing!' And that was the end of my seafaring career – it had been interesting.

EPILOGUE

I remained an engineer surveyor for the next 15 years, working for several different inspection companies after Cornhill, the last one being Bureau Veritas. I am often asked whether I miss the seagoing life, and the answer is that I miss all the good bits. I don't miss having to crawl into roasting hot boilers, or filthy engine crankcases, however, and when I am tucked up in bed at night listening to the wind howling through the trees I think to myself, 'It's a bloody good night not to be bucketing around in the Channel on a Seacat!'

Talking of Seacats, I recently managed to track down Norman Axford, who was one of the other chief engineers with Hoverspeed during my time there. He had remained working on them for almost ten more years, eventually becoming superintendent, and said that things gradually improved over time, as one by one the improvements we had asked for were carried out – but not before many more bump-starting incidents and various other failures. So I don't regret leaving when I did.

You will have gathered that I am something of a steam fan. During my time with Cornhill I did some inspections at the Mid Hants Railway (the Watercress Line) and joined it as a volunteer, eventually becoming a driver in 2004, so my boyhood ambition was finally fulfilled. I also have a 5-inch gauge Britannia Pacific steam locomotive, which took me almost 3,000 hours to build, and which I use to give rides to children at weekends in a local park.

When I finished the Britannia and was wondering what the next project was going to be, my mind went back to the day out on the Stour in Dave Giddy's steam launch, and I decided I would build one of those too. The first thing I did was to join the Steam Boat Association, which helped enormously with sourcing of materials and general good advice. Another very pleasant surprise was meeting up again with Rob Caldwell, who was also a member, and we have been boating together on several occasions since then. I started the project by building a twin-cylinder steam engine, then a three-drum Yarrow-type water-tube boiler, and finally a steam Weirs-type feed pump. The hull I bought in 2014, ready to go, and then I installed all the machinery and plumbing. The maiden steaming was on Bewl Water in 2015, since which time the boat has given me and my family a lot of pleasure at many venues, including Bristol Harbour, Windermere, the Norfolk Broads, the River Medway and the Thames. The boiler has a working pressure of 250 psi – not that I would ever put that much steam through the engine, but many years ago a very good friend of mine gave me a beautiful full-size replica of an A4 locomotive whistle that he had made. At 120–150 psi, which is what most steam launch boilers work at, it sounds very feeble indeed, but at 200 psi and over I can toot away and sound like the world's fastest duck – *Mallard*, in 1938!

In 2018 Astrid and I went on a cruise around the Baltic in *Serenade of the Seas*. I was expecting her to have a diesel-electric drive like most other cruise ships, whose diesel engines drive generators, and everything else, including the propellers, is driven by electric motors – but instead of the diesels our ship had two huge General Electric gas turbines driving a pair of 25 MW generators. The exhaust gases from these went through a waste heat boiler which produced enough steam to drive a steam turbine that generated another 8 MW (10,700 hp) – I was immensely pleased to find that there is still some steam to be found at sea!

APPENDIX 1: EIGHt WAYS to LOSE tHE VACUUM IN tHE MAIN CONDENSER

1. Sea water circulating pump stopped or valves shut.

2. Sea water strainer choked with weed, jellyfish, plastic bags etc.

3. Insufficient gland steam pressure on the turbines, allowing air to enter via the rotor shaft glands.

4. Condenser filling with condensate, which eventually starts to submerge the tubes and reduces the cooling area available. Could be caused by condensate pump failure, problem with condenser level controller or spill valve not opening.

5. Insufficient steam pressure to operate air ejector.

6. Failure to close all cross-connections to auxiliary condenser when shutting it down, thereby allowing air to leak back into the main condenser. Applies also to any shut-down plant with a drain or exhaust lines to the main condenser; if these are not closed then air can leak back.

7. Failing to change over from high to low suction on the main circulating pump as the cargo is discharged – eventually the high suction will draw air and the seawater cooling will fail (JH was most impressed when I thought of this one!).

8. Simply overloading the condenser – quite often when running at full astern. So much steam is being used that it cannot be condensed fast enough. This problem is most likely to occur with high sea temperatures; when it reaches 90°F, for example, the maximum vacuum theoretically possible will be several inches of mercury lower than at 50°F anyway, and the cooling effect of the sea water will of course also be much reduced.

APPENDIX 2: VESSELS I SERVED ON, AND MY RANK

1. SS *Glen Strathallan*. Training ship (converted from deep sea trawler). 690 tons disp. Steam reciprocating (triple-expansion). Engineer cadet.
2. SS *Esso Yorkshire*. Oil Tanker. 94,252 dwt. Steam turbine – 24,000 shp. Engineer cadet.
3. SS *Esso Warwickshire* Oil Tanker. 86,115 dwt. Steam turbine – 24,100 shp. Engineer cadet.
4. SS *Esso Preston*. Coastal tanker. 2,790 dwt. Steam reciprocating (triple-expansion) – 1,350 ihp. Engineer cadet.
5. SS *Esso Edinburgh*. Oil tanker. 47,000 dwt. Steam turbine – 19,000 shp. Engineer cadet/fifth engineer.
6. SS *Esso Durham*. Oil tanker. 40,929 dwt. Steam turbine – 16,000 shp. Fifth engineer.
7. SS *Mobil Daylight*. Oil tanker. 95,715 dwt. Steam turbine – 28,000 shp. Fifth engineer.
8. SS *Mobil Energy*. OBO carrier. 42,648 grt. Steam turbine – 16,270 shp. Fifth/fourth engineer.
9. SS *Mobil Astral*. Oil tanker. 149,269 dwt. Steam turbine – 28,000 shp. Third engineer.
10. MV *Post Runner*. Parcel tanker. 8,847 grt. Slow speed diesel – Gotaverken 8-cylinder, 8,000 ihp. Third engineer.
11. MV *Post Charger*. Parcel tanker. 25,300 dwt. Slow speed diesel – Sulzer 6-cylinder, 12,000 shp. Third engineer.
12. MV *Anco Templar*. Parcel tanker. 23,985 dwt. Slow speed diesel – B&W 7-cylinder, 12,700 ihp. Third engineer.
13. SS *Flinders Bay*. Container ship. 26,756 grt. Steam turbine – 32,450 shp. Third engineer.
14. SS *Jervis Bay*. Container ship. 26,876 grt. Steam turbine – 32,450 shp. Third engineer.
15. MV *Anco Duchess*. Parcel tanker. 17,200 dwt. Slow speed diesel – B&W 6-cylinder, 10,000 ihp. Third engineer.
16. MV *Tiger*. Ro-ro ferry. 3,961 grt. Medium-speed diesel (2) – B&W V10, 5,425 bhp each. Second/chief engineer.

17. MV *Lion*. Ro-ro ferry. 3,987 grt. Medium-speed diesel (2) – Crossley Pielstick V12, 5,175 bhp each. Chief engineer.

18. MV *Panther*. Ro-ro ferry. 3,961 grt. Medium-speed diesel (2) – B&W V10, 5,175 bhp each. Chief engineer.

19. MV *St Magnus*. Ro-ro ferry. 1,205 grt. Medium-speed diesel (2) – MAK 9-cylinder, 2,000 bhp each. Second engineer.

20. MV *Hengist*. Ro-ro ferry. 5,596 grt. Medium-speed diesel (2) – Crossley Pielstick V16, 7,400 bhp each. Second/chief engineer.

21. MV *Horsa*. Ro-ro ferry. 5,590 grt. Medium-speed diesel (2) – Crossley Pielstick V16, 7,400 bhp each. Second/chief engineer.

22. MV *Vortigern*. Ro-ro ferry. 4,371 grt. Medium-speed diesel (2) – Pielstick V16, 7,250 bhp each. Second engineer.

23. MV *Cambridge Ferry*. Train Ferry. 3,294 grt. Medium-speed diesel (2) – Mirrlees 7-cylinder, 1,835 bhp each. Second/chief engineer.

24. PS *Kingswear Castle*. Paddle steamer. Compound diagonal paddle engine. Served three days as volunteer chief engineer.

25. MV *St Christopher*. Ro-ro ferry. 7,399 grt. Medium-speed diesel (2) – Pielstick V16, 10,250 bhp each. Second engineer.

26. MV *Seafreight Freeway*. Ro-ro ferry. 5,739 grt. Medium-speed diesel (2) – MWM V12, 5,900 bhp each. Second/chief engineer.

27. MV *Hounslow*. Sludge Carrier. 2,132 grt. Medium-speed diesel (2) – Napier 6-cylinder, 1,375 bhp each. Chief engineer.

28. MV *Bexley*. Sludge Carrier. 2,132 grt. Medium-speed diesel (2) – Napier 6-cylinder 1,375 bhp each. Chief engineer.

29. MV *Newham*. Sludge Carrier. 2,132 grt. Medium-speed diesel (2) – Napier 6-cylinder, 1,375 bhp each. Chief engineer.

30. MV *Thames*. Sludge carrier. 2,663 grt. Medium-speed diesel – Mirrlees 6-cylinder, 2,500 bhp. chief engineer

31. MV *Stena Fantasia*. Ro-ro ferry. 25,122 grt. Slow speed diesel (2) – Sulzer 7-cylinder, 8,725 bhp each. Second engineer.

32. MV *Stena Invicta*. Ro-ro ferry. 19,763 grt. Slow speed diesel (2) – B&W 8-cylinder, 8,480 bhp each. Second/chief engineer.

33. MV *Pride of Winchester*. Ro-ro ferry. 6,386 grt. Medium-speed diesel (3) – Werkspoor 8/9*-cylinder, 14,283 bhp total. * This ship had port and starboard engines of 8 cylinders each and a centre engine of 9 cylinders. Third engineer.

And last but not least the Seacats. On all of these I sailed as chief engineer:

34. *Hoverspeed France* (Incat 026)
35. *Hoverspeed Boulogne* (Incat 027)
36. *Seacat Scotland*. (Incat 028)
37. *Hoverspeed Great Britain*. (Incat 025)

They were all 3,003 grt and powered by four Ruston RK270 medium-speed diesels of 5,000 bhp each, driving Lips water jets.